Acclaim for
THE BOOK OF HUMANS

An Amazon Best Science Book of the Year

"Rutherford describes [*The Book of Humans*] as being about the paradox of how our evolutionary journey turned 'an otherwise average ape' into one capable of creating complex tools, art, music, science, and engineering. It's an intriguing question, one his book sets against descriptions of the infinitely amusing strategies and antics of a dizzying array of animals."
—The New York Times Book Review

"A smooth, expert, and often startling history that emphasizes that no behavior separates us from other animals, but we remain an utterly unique species."**—Kirkus Reviews**

"A refreshing and perspective-altering view of the complex history of life on Earth."**—Publishers Weekly**

"Rutherford is an engaging, witty writer. . . . Entertaining."
—The Guardian

"A kind of intellectual enema, exposing the popular myths about human exceptionalism."**—New Scientist**

"Recommended."**—Scientific American**

"I love learning new, surprising facts, the kind that make children say 'Did you know?' Did you know there are Australian hawks who pick up burning sticks and transport them to start a fire, where they then eat roasted animals? . . . Highly stimulating, lots to think about, lots to learn. Very well-written . . . thoroughly recommended."
—Richard Dawkins

"Fascinating . . . enlightening . . . Rutherford writes with clarity, authority and humor. His research is thorough and so current that most readers will be wowed."**—BookPage**

"Engaging, entertaining, and highly accessible."—*Choice*

"[Rutherford] is an amiable guide to this look at the ways in which we are members of the animal kingdom yet unequivocally at the head of the class, creature-wise."—*Toronto Star*

"An energetic exploration of the animal kingdom reveals what humans share with other creatures, and what makes us different."—*Shelf Awareness*

"Engaging, accessible, and highly recommended."
—*Library Journal*

"[Rutherford] writes with intellectual authority and also, as a popular lecturer and broadcaster, expresses himself in a clear and persuasive manner with natural charm."
—*The Spectator*

"I loved this book. An outstandingly clear and witty account that shows beyond doubt how much we are part of the animal world, and yet at the same time how different we have become."
—**Henry Marsh**, author of *Do No Harm*

"This delightful and charming book will change the way you see yourself and your place in the natural world."
—**Ed Yong**, author of *I Contain Multitudes*

"Adam Rutherford is a master storyteller. *The Book of Humans* is packed to the brim with intriguing tales, clever twists, and up-to-the-minute scientific discoveries, offering a completely new perspective on who we are and how we came to be."
—**Hannah Fry**, author of *Hello World*

"Charming, compelling, and packed with information. I learned more about biology from this short book than I did from years of science lessons."
—**Peter Frankopan**, author of *The Silk Roads*

"If teaching is what makes humans special, then Adam Rutherford is superhuman—the paragon of teachers, a truly gifted transmitter of knowledge: lucid, enlightening, witty, and delightful."
—**Kate Fox**, codirector of the Social Issues Research Centre

"Adam Rutherford is a superb communicator who eruditely explores the borderlands of history, archaeology, genetics, and anthropology in this fascinating tour of our species."
—**Dan Snow**, host of the podcast *Dan Snow's History Hit*

"Rutherford is a delightful, eclectic, hilarious, and often filthy guide to what we know about human genetics. But even more than that, there are parts of this book I wish were graven into the walls of public buildings and above the desk of anyone who writes about science. Next time someone tells you that men 'evolved' to behave in one way while women 'evolved' to behave differently, pull out your copy of *The Book of Humans* and set them right."
—**Naomi Alderman**, author of *The Power*

Acclaim for
ADAM RUTHERFORD

"An enthusiastic guide and a good storyteller."
—*The Wall Street Journal*

"Rutherford unpeels the science with elegance."　　—*Nature*

"With intellectual authority [. . . Rutherford] expresses himself in a clear and persuasive manner with natural charm."
—*The Spectator*

"Rutherford is a bold, confident storyteller."　　—*Genome*

Also by Adam Rutherford

How to Argue With a Racist:
What Our Genes Do (and Don't) Say About Human Difference

A Brief History of Everyone Who Ever Lived:
The Human Story Retold Through Our Genes

Creation: How Science Is Reinventing Life Itself

THE
BOOK OF
HUMANS

A BRIEF HISTORY OF CULTURE, SEX, WAR, AND THE EVOLUTION OF US

ADAM RUTHERFORD

Illustrations by Alice Roberts

THE EXPERIMENT

NEW YORK

THE BOOK OF HUMANS:
A Brief History of Culture, Sex, War, and the Evolution of Us
Copyright © 2018, 2019, 2020 by Adam Rutherford
Illustrations copyright © 2018, 2019, 2020 by Alice Roberts

Originally published in Great Britain by Weidenfeld & Nicolson, an imprint of
the Orion Publishing Group Ltd., a Hachette UK company, in 2018.
First published in North America in revised form as *Humanimal* by
The Experiment, LLC, in 2019. This paperback edition first published in 2020.

The Experiment, LLC
220 East 23rd Street, Suite 600
New York, NY 10010-4658
theexperimentpublishing.com

THE EXPERIMENT and its colophon are registered trademarks of The
Experiment, LLC. Many of the designations used by manufacturers and sellers to
distinguish their products are claimed as trademarks. Where those designations
appear in this book and The Experiment was aware of a trademark claim, the
designations have been capitalized.

The Experiment's books are available at special discounts when purchased in bulk
for premiums and sales promotions as well as for fund-raising or educational use.
For details, contact us at info@theexperimentpublishing.com.

Library of Congress Cataloging-in-Publication Data available upon request

ISBN 978-1-61519-590-9
Ebook ISBN 978-1-61519-532-9

Cover and text design by Beth Bugler
Cover illustrations by Ksanawo/Shutterstock.com
Author photograph by Stefan Jakubowski

Manufactured in the United States of America

First paperback printing April 2020
10 9 8 7 6 5 4 3 2 1

CONTENTS

PART TWO: THE PARAGON OF ANIMALS

ILLUSTRATIONS
by Alice Roberts

Introduction

"What a piece of work is a man!" marvels Hamlet, in awe at our specialness.

> How noble in reason! How infinite in faculty!
> In form, in moving, how express and admirable! In action
> how like an angel!
> In apprehension how like a god! The beauty of the world!
> The paragon of animals!

"The paragon of animals" is a lovely phrase. Hamlet exalts us as truly special, touching the divine, limitless in our thought. It's a prescient phrase, too, as he raises us above other animals while acknowledging that we are one. Just over 250 years after William Shakespeare wrote those words, Charles Darwin irrefutably cemented humankind's classification as an animal—the slightest of twigs on a single, bewildering family tree that encompasses four billion years, a lot of twists and turns, and a billion species. All of those organisms—including us—are rooted in a single origin, with a common code that underwrites our existence. The molecules of life are universally shared, the mechanisms by which we got here the same: genes, DNA, proteins, metabolism, natural selection, evolution.

Hamlet then ponders the paradox at the heart of humankind:

What is this quintessence of dust?

We are special, but we are also merely matter. We are animals, yet we behave like gods. Darwin sounds a bit like Hamlet, declaring that we have "god-like intellect," yet we cannot deny that man—and, to bring his language into the twenty-first century, woman—carries the "indelible stamp of his lowly origin."

This idea, that humans are special animals, is at the root of who we are. What are the faculties and actions that put us on a pedestal above our evolutionary cousins? What makes us animals, and what makes us their paragon? All organisms are necessarily unique, so that they can exist within and exploit their own unique environment. We certainly think of ourselves as pretty exceptional, but are we really more special than other animals?

Alongside Hamlet and Darwin comes a possible challenge to our ideas of human exceptionalism, from an arguably lesser piece of modern culture, the animated superhero film *The Incredibles*: "Everyone is special . . . which is another way of saying that no one is."

Humans *are* animals. Our DNA is no different from anything that has lived in the last 4,000 million years. The coding system employed within that DNA is no different, either; the genetic code is universal as far as we know. The four coded letters that make up DNA (known as A, C, T, and G) are the same in bacteria and bonobos, orchids, oaks, bedbugs, barnacles, triceratops, *Tyrannosaurus rex*, eagles, egrets, yeast, slime molds, and ceps. The way they are arranged in those organisms, and how they are translated into the protein molecules that enact the functions of a living being, are all fundamentally the same, too. The fact that life is organized into discrete cells is also universal,* and these incalculably numerous cells

---

* Viruses are normally and traditionally excused from this definition; arguments rage over whether viruses are living or not, though I vacillate between

harvest energy from the rest of the universe in a process common to all of them.

These principles are three of the four pillars of biology: universal genetics, cell theory, and chemiosmosis, which is a rather technical yet elegant word for the basic process of cellular metabolism—how cells draw energy from their surroundings, to be spent in the process of living. The fourth pillar is evolution by natural selection. Combined, these grand unifying theories coalesce to reveal something unarguable—that all life on Earth is related by common ancestry, and that includes us.

Evolution is slow, and Earth has been host to life for the vast majority of our planet's existence. The timescales we talk about so casually in science are utterly baffling to comprehend. Even though we are a latecomer to life on Earth, our species is more than 3,000 centuries old. We have traversed that ocean of time largely unchanged. Physically, our bodies are not drastically different from *Homo sapiens* in Africa 200,000 years ago.* We were physically capable of speaking then as we do today, and our brains were not significantly different in size. Our genes have responded in small part to changes in the environment and diets as we migrated within and out of Africa, and genetic variants account for the minuscule percentage of DNA that spells out the differences between individuals, changes in the most superficial characteristics—skin color, hair texture, and a few others. But if you tidied up a *Homo sapiens*

_____

not caring and thinking that for all intents and useful purposes they display the characteristics of being alive. That they cannot reproduce themselves without the presence of a cellular living entity is, to my mind, not relevant. No organism has ever existed without dependence on another. The role of viruses in evolution cannot be overstated and has been a major driver of the continuation of life for its entire duration, as is discussed later.

* The earliest *Homo sapiens* are found in Morocco and are around 300,000 years old, but these are sometimes referred to as archaic, rather than anatomically modern, humans, the oldest of which are more like 200,000 years old.

woman or man from 200,000 years ago, gave them a haircut, and dressed them in twenty-first-century clothes, they would not look out of place in any city in the world today.

There's a conundrum in that stasis. Though we may not look different, humans did change, and profoundly. There's debate about when this transition occurred, but by 45,000 years ago, something had happened. Many scientists think that it was a sudden change—sudden in evolutionary terms means hundreds of generations and dozens of centuries, rather than a thunderbolt. We don't quite have the language to relate to the timescales involved in such transitions. But what we can observe from the archaeological record is that we see the emergence and accumulation of a number of behaviors that are associated with modern humans, and there was a time before that where we see fewer or none of them. Given how long life has existed on Earth, this switch happened relatively in a heartbeat.

The transformation occurred not in our bodies or physiology or even in our DNA. What changed was culture. In scientific terms, culture refers broadly to the artifacts that are associated with a particular time and place. They include things like tools, blade technology, fishing gear, and use of pigment for decorative purposes or jewelry. The archaeological remains of a hearth show the ability to control fire, to cook, and maybe its position as a social hub. From material culture, we can infer behavior. From fossils we can try to piece together what people looked like, but with archaeological evidence of the paraphernalia of our ancestors' lives, we can address what people *were like* in prehistory, and when they became like that.

By 40,000 years ago, we were designing decorative jewelry and musical instruments. Symbolism in our art was rife, and we were inventing new weapons and hunting technology. Within a few millennia, we had fostered dogs into our lives—tamed wolves that accompanied our search for food long before they became our pets.

The concatenation of these behaviors is sometimes referred to as the Great Leap Forward, as we jumped into a state of intellectual sophistication that we see in ourselves today. Alternatively, it's a "cognitive revolution," but I dislike the use of that phrase to describe a process that was both continuous and probably lasted a few thousand years or more—real revolutions should be thunderbolts. Nevertheless, modern behavior emerges permanently and quickly in several locations around the world. We began to carve complex figurines, both realistic and abstract, sculpted make-believe chimeras out of ivory, and we adorned cave walls with pictures of hunts and of animals important to our lives. One of the two earliest known pieces of figurative art by *Homo sapiens* that we know of is a 40,000-year-old twelve-inch statue of a lean man with the head of a lion. It was carved from a mammoth tusk during the last Ice Age.

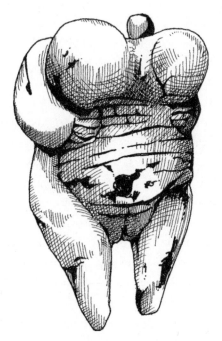

The Venus of Hohle Fels

Soon after that time we were making small statues of women. They are known today as Venus figurines. We don't know if there was a specific purpose to these dolls, though some researchers think they may have been fertility amulets, as their sexual anatomy is exaggerated: bosomy women with swollen labia, and often bizarrely small heads. Maybe they were just art for art's sake, or toys. Either way, to create such sculptures requires great skill and foresight, and a capacity for abstract thought. A lion-man is an imagined being. The Venus amulets are deliberate misrepresentations, abstractions of human bodies. These figures cannot exist in isolation, either; artisan craft requires practice, and though today only a handful of these beautiful works of art remain, they must represent an iterative process, a lineage of skilled craftspeople.

Some of these types of traits pop up before the full transition to our modern behavior, but they do so fleetingly, and then vanish from the archaeological record. *Homo sapiens* were not the only humans to have existed in the last 200,000 years, and not the only ones to have refined culture. *Homo neanderthalensis*, far from the brutes of popular lore, were simply people, too. We are wrong to think of them as merely upright apes, living in dust with crude language and tools, all set for extinction. Neanderthals showed clear signs of modern behavior: They made jewelry, employed complex hunting techniques, used tools, had a control of fire, and made abstract art. We have to consider that they also were sophisticated in a way indistinguishable from our direct

Homo sapiens ancestors, which undermines the uniqueness of our own forward leap. Though we have traditionally considered Neanderthals to be cousins to us, they were also ancestors: We now know that our lineage and theirs diverged more than half a million years ago, and both groups were isolated in time and space for almost all of that period. But our ancestors left Africa some 80,000 years ago, and were immigrants into Neanderthal territory.

We reached Europe and central Asia, and around 50,000 years ago, we bred with them. Their bodies were different enough that they lie outside the range of human physical diversity as we see it today—a bit less chin, a bit more chest, heavy-set brows, and robust faces. They weren't so different that we didn't have sex with them, women and men from both sides of the species fence, and together we had children. We know this because our genes are in their bones, and theirs are in our living cells. Most people of European descent carry a small but significant percentage of DNA that was acquired from Neanderthals, and this blurs any hope of a clear boundary between two sets of people that we have declared separate species—that is, organisms that cannot produce fertile offspring. Though Neanderthal DNA is slowly being purged from our genomes for reasons that are not fully understood, humans today bear their living genetic heritage, as we do the genes of another type of human, the Denisovans, farther to the east, and maybe others that are yet to be discovered but whose legacy sits within our DNA.

When we first met, the Neanderthals and those other people were not long for this world, and by around 40,000 years ago, *Homo sapiens* had outlived the last of them. Whether the Neanderthals had undergone a full transition to the behavioral modernity as we saw in *Homo sapiens*, we do not know, and may never know, but the evidence is pointing toward those cavemen and women being much like ourselves in every way.

We lived and they died. We don't know what gave *Homo sapiens* the edge over Neanderthals. All life is set for extinction over a long enough timescale; more than 97 percent of species that have ever existed are already gone. The Neanderthals' tenure on Earth was much longer than we have racked up so far, and we are yet to firmly understand why their light was finally snuffed out 40,000 years ago. We don't think there were ever very many Neanderthals, which may have contributed to their demise. Maybe we outsmarted them. Maybe we brought with us diseases that we had lived with and earned immunity to, but which were lethal to a virgin population. Maybe they simply petered out of existence. What we do know though is that around this time, the last type of human began to permanently and globally show signs of who we are today.

We certainly outbred all our nearest relatives. *Homo sapiens* went forth and multiplied very effectively. We're the dominant life form on Earth by many measures, if ranking matters to you (though bacteria outnumber us—you carry more bacterial cells than human ones—and are far more successful in terms of longevity. They have a four-billion-year lead on us, and no prospect of extinction). Today there are upwards of seven billion humans alive, more than at any time in history, and that number is still rising. Through our ingenuity, science, and culture, we have eradicated many diseases, drastically reduced infant mortality, and extended lifespan by decades.

Hamlet marvels at our brilliance, as have scientists, philosophers, and religions for millennia. But the progress of knowledge has chipped away at our specialness. Nicolaus Copernicus dragged us away from a world at the center of the universe to one merely orbiting an ordinary star. Twentieth-century astrophysics revealed our solar system as an average one among billions in our galaxy, which itself is one of billions in the universe. We still only know of one world that harbors life, but since 1997, when the first planets beyond our Sun's gravity were discovered, we have learned of thousands in the heavenly firmament, and in April 2018 a new satellite was launched specifically to seek out strange new worlds. We're getting a good grip on the conditions required for chemistry to transition into biology, and for life to emerge from a sterile rock. The question of whether there is life beyond Earth has mutated: It would now be surprising if there *weren't* living things elsewhere in the universe. That is all still to come, and for now, we only know of life on Earth. But we might not be as unique as we once thought, and the more we learn the clearer that becomes.

On Earth, Charles Darwin began the process of inching us back into the natural world, and away from special creation. He showed that we are animals, evolved from other animals, and placed us firmly as a creature begotten not created. All of the incontrovertible molecular evidence of those pillars of biology was yet to come when he exposed the world to his big idea in 1859 in *On the Origin of Species*. He avoided including humans in that great work, but teased us that his mechanism of natural selection would soon shed light upon our own origins. In *The Descent of Man* in 1871, he applied his meticulous and foresightful brain to our genesis, and cast us as an animal evolved just like every organism in Earth's history. Mostly bald, you're an ape, descended from apes, your features and actions carved or winnowed by natural selection.

In that sense we are not special. We evolved with a biology indistinct from all life, and under the auspices of a mechanism that is similarly universal. But evolution also equipped us with a suite of cognitive powers that gave us, ironically, a sense of separateness from nature, because it allowed us to develop and refine our culture to a level of complexity well beyond any other species. It gave us a clear sense that we are special, and specially created.

Yet many of the things once thought to be uniquely human are not. We have extended our reach so far beyond our grasp by utilizing nature and inventing technology. But many animals also use tools. We have decoupled sex from reproduction, and almost always have sex for fun. Scientists are reluctant to admit the possibility of pleasure in animals, but even so, a huge proportion of sexual activity in animals does not and cannot result in reproduction. We are frequently a homosexual species. Once—and in many places to this day—homosexuality was decried as *contra naturam*, a crime against nature. In fact, sexual acts between members of the same sex abound in nature, in thousands of animals, and, for example, may well dominate male giraffe sexual encounters.

Our ability to communicate appears to trump all other animals, though maybe we just don't know what they're saying yet. I am writing this book and you are reading it, which is a degree of communication that has evolved far beyond any other level we have observed in any other species. Though that certainly makes us different, a mantis shrimp doesn't give two hoots about that. It can see in sixteen different wavelengths of light compared to our puny three,* which is rather more useful to them than all the culture and self-regard that we have mustered over the millennia.

Nevertheless, a book is a thing that typifies the gap between us and all other beasts. It is the sharing of information generated by thousands of others, almost none of whom I am closely related to. I have studied their ideas and recorded them into a tool of almost unimaginable complexity, so that our minds might be enriched with this collection of stories that are new and hopefully interesting to anyone who cares to pick it up.

This is a book about the paradox of how we became us. It is an exploration of an evolution that bestowed enormous powers of intellect on an otherwise average ape, to create tools, art, music, science, and engineering. Through old bones and, nowadays, genetics, we know about the mechanics of our evolutionary journey through the eons (though there is so much still to discover), but we know far less about the development of our behavior, of

---

* Or four: We are beginning to think that some women are tetrachromats, meaning that they have photoreceptors that are optimized to detect four primary colors, rather than the standard trichromatic three. The new primary color will be in the green range.

our minds, and of the way that we uniquely evolved into the cultural and social beings that we are today.

At the same time, though, it is a book about animals, of which we are one. We're a self-centered species, and we find it hard not to see ourselves and our behaviors in other animals. Sometimes those characteristics do have a shared origin with our own. Often they do not. Regardless of their genesis, I am attempting to demystify our own behavior by pointing to where else on Earth we see those traits, and trying to sort the things that are uniquely us, shared with close evolutionary cousins, or just things that look similar, but are in fact unrelated. I'll be examining the evolution of technology in humans—having mastered the crafting of stones, and sticks, and fire hundreds of thousands of years ago—and in the many other animals that also use tools. Evolutionary biologists love thinking about sex, and I'll be delving in, not only to try to understand how we decoupled sex in all its myriad forms from reproduction, but how the sex lives of animals are also a carnival of delights that are not always simply the direct manifestation of the biological imperative to create offspring. While this is a celebration of both us and the wondrous variation in nature, we are indubitably a creature capable of less than angelic behavior, of creating horrible nightmares—violence, warfare, genocide, murder, rape. Are these different from the often horrifying behaviors that are part of the brutal natural world, the violence and sexual practices that don't get showcased on television documentaries? In the final part, I will be scrutinizing the reasons behind the evolution of behavioral modernity—meaning the emergence of people who are like us today. Our bodies became modern long before our minds did, which is a puzzle worth examining.

Biologists appraise the wonders of evolution, sometimes to understand ourselves, often to understand the grand scheme of life on Earth. This book is a glimpse of the epic meandering journey that every organism has made. After all, we are the only ones who can appreciate it.

What a piece of work we are!

The pillars of biology are firmly in place, installed over the last two centuries and tested over and over again. We have bound the principles of natural selection to genetics, in cells powered by chemistry. We have aligned these principles in history, to draw a picture of how life spread from such a simple beginning in the basement of the oceans to every inch of this planet. You might think that this means the study of life on Earth is pretty much done, and now we're just filling in the details. But science never sleeps, because there are always leviathan gaps in our knowledge. Most of nature remains unobserved, and it continues to utterly astound us with new discoveries every day, new species, and new traits in animals and other organisms that we simply have neither seen before nor perhaps even conceived of.

Some of the things described in the pages that follow were only discovered in 2018, the year I finished writing this book. That may mean details are scant, or have been seen only once or on a few occasions. It may mean these newly observed behaviors are outliers, truly unusual characteristics. Others might be generalizable to many species, or even all. Some may turn out to be not what we originally thought. For all the glorious documentaries that we see on television, most animals spend almost all of their lives unseen by human eyes, and live in environments that are inhospitable or alien to us. That is the nature of science: Seek and ye shall find. Studying these animals is important on its own terms, and may yet provide insight into our own condition.

Sometimes these behaviors appear to have a shared evolutionary origin with us. Others exist in non-human animals because they are clearly of great use in the struggle for existence, and have evolved many times over, just as insects, bats, and birds all have wings but with little in common in their histories of acquiring flight. The philosopher Daniel Dennett calls these "good tricks," meaning that they are characteristics of such benefit that they arise many times in history. Flight is a good trick, and has evolved repeatedly in distantly related creatures, but it has also evolved many times over within the same groups of creatures. Evolution can be efficient in that way; once there is a plan to make a particular trait, that plan can be deployed whenever desirable. Insect wings have come and gone dozens, maybe hundreds, of times in the last few hundred million years to suit survival in the local environment, though the genetic mechanism that underlies wings remains largely unchanged during this time. Flying is only useful when it's useful, and it's a costly activity, so can be discarded, and the genes filed away, when not needed, like a winter wardrobe.

There are plenty of potential pitfalls in studying our own evolution. Just as we must be careful about ascribing similarity of function with common origins, we must also be cautious about confusing our behavior today with a presumption that that is why the behavior emerged in the first place. There are many tempting myths about the origin of our bodies and behaviors that teeter near the edge of pseudoscience. Let me be clear on this: all life is evolved. But that doesn't necessarily mean that all behaviors are explained with the central idea of evolution, which is adaptation. Many behaviors, especially in us, are there as by-products of our evolved existence, and not because they have specific functions that aid our survival. This fallacy is particularly prevalent in our sexual behaviors, which we will inspect in detail. We see familiar sexual behaviors in animals, some of which are associated with pleasure in us, and some with criminal violence. No matter how neat or appealing an explanation might be, science looks to facts and evidence, and an ability to test an idea to destruction.

Every evolutionary pathway is unique, and while all living beings are related, how each one came to be is a different story, with different pressures driving selection, and random changes in DNA providing the template from which variation, selection and evolutionary change can occur. Evolution is blind, mutation is random, selection is not.

Error and trial is a conservative process; radical biological change normally results in death. Some evolutionary developments are clearly so useful that they never truly go away. Vision is one example. Being able to see in the oceans clearly conferred a significant advantage for whatever life-form first acquired vision, more than 540 million years ago—you can see things you wish to eat and move toward them, you can see things that wish to eat you and swim away. Once it had evolved, vision spread rapidly. Since then, the genetic program for phototransduction—that is, converting light into sight—has remained virtually identical in all organisms that can see. In contrast, a crow with a bent stick wheedling out a fat grub from the bark on a tree is a skill that has evolved entirely independently of a chimpanzee doing exactly the same thing, and has little specific genetic underpinning in common. All abilities are evolved, which doesn't mean that they all have common roots. Unpicking and filtering the similarities and difference in behaviors that appear familiar to us is crucial in understanding our own evolution.

We have to separate out all of the attributes discussed in this book, even though each is dependent on others. We cannot recreate the order or circumstances in which they appeared. Our brains expanded, our bodies changed, our skills sharpened and we socialized differently. We ignited sparks and lit fires, tilled the earth, crafted myths, created gods and commanded animals. The beginning of culture relied upon all of these things, powered by the flow of information and expertise. It was not an apple that gave us this knowledge—apples are a product of our own agricultural ingenuity. It was how we lived our lives. We began living in populations that grew to sizes where kin became communities, and tasks within communities fell to specialists—musicians, artists, craftspeople, hunters, cooks. In the transfer of the wisdom of these experts—in the interconnectedness of minds—modernity arose. Uniquely, we accumulate culture and teach it to others. We transmit information, not just via DNA down the generations, but in every direction, to people with whom we have no immediate biological ties. We log our knowledge and experiences, and share them. It is in the teaching of others, the shaping of culture, and the telling of stories, that we created ourselves.

Darwin, with typical prescience, suspected that this might be the case:

> Man alone is capable of progressive improvement. That he is capable of incomparably greater and more rapid improvement than is any other animal, admits of no dispute; and this is mainly due to his power of speaking and handing down his acquired knowledge.

Crucially, we are the only species to have held ourselves up to the light, to have asked, "Am I special?" Paradoxically, the answer turned out to be both no and yes.

Over the eons, we have moved from being not particularly special animals, to thinking ourselves uniquely created and distinct from the rest of the living world, to a sort of quantum state where we can occupy both positions at the same time. Here is a compendium of what unequivocally fixes us within the animal kingdom, and simultaneously reveals how we are extraordinary.

PART ONE

Humans and Other Animals

Tools

Humans are creatures imbued with technology. That is a word that has taken on a specific meaning in the modern age. I write these words on a computer, with an internet browser on in the background connected via Wi-Fi. We tend to think of these types of electrical gadgets and services as being the embodiment of technology today. The science-fiction writer Douglas Adams came up with three rules concerning our interaction with technology:

1. Anything that is in the world when you're born is normal and ordinary and is just a natural part of the way the world works.
2. Anything that's invented between when you're fifteen and thirty-five is new and exciting and revolutionary and you can probably get a career in it.
3. Anything invented after you're thirty-five is against the natural order of things.

Certainly, there is a seemingly constant suspicion of new technologies in the media, especially by older people expressing concern for the young: *won't somebody think of the children?*

It is the same as it ever was. In the fifth century BCE, Socrates railed against the dangers of a new disruptive technology for fear that in young men it would nurture:

Forgetfulness in the learners' souls, because they will not use
their memories . . . they will be hearers of many things and
will have learned nothing; they will appear to be omniscient
and will generally know nothing; they will be tiresome company, having the show of wisdom without the reality.

The technology that prompted Socrates's ire was writing. Two
thousand years later, a sixteenth-century Swiss polymath, philolo-
gist and scientist called Conrad Gessner expressed similar angst at
the potential of another innovation for information technology—
the printing press.

Plus ça change . . . The current cultural techno-malaise is born of
our time spent interacting with screens. The media, both in print
and online, endlessly fret about the amount of time we spend in
front of screens, and the potential damage that it might cause.
Everything from low-level delinquency to spree killing to autism
to schizophrenia has been attributed to excessive screen time in
recent years. It's a frustrating pseudo-scientific discussion in general, as the terms of the problem are poorly delineated and
ill-defined. Is five hours immersed in a video game alone of equal
impact to five hours absorbing a book on a reading device? Does it
matter if the game features violence or puzzles or both, or if the
book is an incitement to violence or to manufacture weapons? Is
watching a film at the cinema the same as playing a video game
with your family?

The data is not yet available, and studies that have been done so
far have not drawn any strong conclusions or have been flawed in
some way or other. Part of the discourse though is that we spend too
much time on screens, when we should be doing more creative or
cultural things or expressing ourselves without a reliance on technology. Of course, a paintbrush is a technological tool, as is a pencil,
a sharpened stick or a particle accelerator. Very nearly nothing we

do, artistic, creative or obviously scientific, could exist without technology underpinning it. Singing, dancing, even some forms of athletics and swimming, operate without direct reliance on an external technology, but as I watch my daughter tie her hair into a bun and spray it into place, clip her battered toenails and don her pointe shoes before ballet I can't help but think how we are an animal whose entire culture and existence is completely dependent on tools.

So, what is a tool? There are a few definitions. Here is one from a key textbook on animal behavior:

> The external employment of an unattached or manipulable attached environmental object to alter more efficiently the form, position, or condition of another object, another organism, or the user itself, when the user holds and directly manipulates the tool during or prior to use and is responsible for the proper and effective orientation of the tool.

Which is wordy but covers most bases.* Some definitions make a distinction between use of a found object and a modified item, which qualifies it as technology. The key idea is that a tool is a thing external to the animal's body that is used to exert a physical action for the animal that extends its powers.

Tools are an inherent part of our culture. Sometimes we talk about cultural evolution in opposition to biological evolution, the former being taught and passed down socially, the latter being encoded in our DNA. But the truth is that they are intrinsically linked, and a better way to think about it is as gene-culture co-evolution. Each drives the other, and cultural transmission of ideas and skills requires a biologically encoded ability to do so. Biology enables culture, culture changes biology.

---

* *Animal Tool Behavior: The Use and Manufacture of Tools by Animals* by Robert W. Shumaker, Kristina R. Walkup and Benjamin B. Beck (Johns Hopkins University Press, 2011).

Millions of years before the invention of the digital watch, we had an obligate technological culture. We have even specifically acknowledged our technological commitment in scientific nomenclature. One of our earliest genus cousins–probable ancestors—is named *Homo habilis*. Literally, this means "handy man." They were a people that lived between 2.1 and 1.5 million years ago in east Africa. There are a few specimens that have been classified as *habilis*, which generally have flatter faces than the earlier Australopithecines from around three million years ago, but still retain long arms and small heads—their brains were typically half the size of ours. To look at, *Homo habilis* would have been more ape-man than man-ape. They were probably the ancestors of the more gracile *Homo erectus*, though coexisted with them as well, maybe indicating that *Homo habilis* diverged within its own species group.

Their handyman status is largely due to the discovery of specimens surrounded by evidence of lithic—that is, stone—technology. Some researchers suppose that the presence of tools represents the boundary between the genus *Homo* and what came before, meaning that humans are actually defined by tool use. The densest collections associated with *Homo habilis* come from the Olduvai Gorge in Tanzania, and this type of technology is referred to as the Oldowan tool set. There is a lot of technical jargon involved in describing this kit and how they were worked; "lithic reduction" is one such term, which broadly means chipping a stone, often quartz, basalt or obsidian, to shape and sharpen. Many of the archaeological clues come in the form of lithic flakes—the detritus knapped from a raw stone into a tool, when the tool itself is lost in time. Obsidian* is an igneous rock—a type of volcanic glass, and a good choice for a cutting tool, as it forms edges so sharp that some surgeons use it today in preference to steel scalpels.

---

* Geologists have all the best names: obsidian is a rock formed when felsic lava rapidly cools on the edges of rhyolitic flows; that means it's rich in the silicate compounds feldspar and quartz.

These actions imply a cognitive ability that enables selection of suitable stones, and a plan. You need a hammer stone and a platform, an anvil, on which to chip away at the raw material. Knapping is a deliberate and skillful activity, and the set required contains different tools. Some are heavy duty, such as the Oldowan chopper, which we think was used as an ax head. Others are lighter-duty tools—scrapers for removing meat from skins, chisel-edged stones called burins and other tools for engraving wood. Again, this variation in the overall set of tools presupposes a cognitive ability to distinguish appropriate tools for different actions.

Homo habilis is among the earliest members of the lineage that we have decided is human, and tool use is part of that definition. But this artificial boundary has not been borne out in scientific history; handy man wasn't the first to get handy. A thousand kilometers to the north of Olduvai is Lomekwi, on the western shore of Lake Turkana, another of the key areas in the nursery of early humans. This was the site of the discovery in 1998 of a specimen that has been designated *Kenyanthropus platyops*, roughly meaning "flat-faced Kenyan man."* It's a not-uncontroversial earlier great ape, that some have argued is morphologically similar enough to Australopithecus to suggest that it is not a separate species. I'm not sure it matters that much, as our taxonomic definitions are blurred at these arbitrary boundaries, and many assumptions must be made due to the specimens being few and far between—fragments from

---

* Historically, the word "man" has been used to describe these species in common parlance, as in Neanderthal man, Cro-Magnon Man, etc. It's a casual usage that is annoying in that it fails to recognize 50 percent of our species but it can be easily corrected by generally using "human," as in humankind, which is an easy and inclusive fit. In this case though, "human" specifically refers to the genus *Homo*, which *Kenyanthropus platyops* is not in, but anthropus implies humanness, though in Greek it literally means "man," so I'm not quite sure how to represent it here. It is a hominin, which includes both *Homo*—the humans—and Australopithecines, whose name roughly translates as "southern ape-like things."

more than 300 Australopithecine individuals have been found, but only one *Kenyanthropus*.

In 2015, a wandering team of researchers from Stony Brook University in New York took a wrong turn in Lomekwi and stumbled into a site scattered on the surface with lithic detritus indicative of intentional tool-making. After excavating further, they found many other fragments and tools themselves. The dirt in which they were found could be accurately dated, which is not always easy, but in this case, was dependent on layers of volcanic ash, and the geological phenomenon of magnetic pole reversal.* The tools found are not quite as sophisticated as the Oldowan set, but are much older, probably 3.3 million years. In one case a lithic flake could be matched to the stone from which it was chipped. It is viscerally powerful: imagine an ape-like person sat right there, intentionally chipping at a rock with purpose in mind. Maybe he or she wasn't happy with how it split, discarded both halves, and moved onto something else. Or maybe it was chased away by a ravenous predator. There, the pieces lay undisturbed for more than three million years.

We don't know who it was who sat and carved those tools, though we do know that it was a creature that pre-dates the origin of the genus *Homo*—the humans—by maybe 700,000 years, and may well be the flat-faced Kenyans. The Oldowan tool set has now been found in key sites all over Africa where other significant evidence of human

---

* The magnetic poles are constantly moving and have flipped many times in our planet's history. We're not sure why and cannot predict when they will flip. The change happens over thousands of years, and no pattern has yet been discovered for the known times that magnetic north and south have reversed. But these reversals are recorded in microscopic fragments in rocks, and thus are useful for dating when the rocks were formed. The North Pole is currently moving south at a rate of about a few miles a year, though this is nothing to be worried about—it is too slow to have any noticeable effect on us, or the many migratory animals that have magnetoception and navigate using Earth's natural polarity.

presence is known, including Koobi Fora on the east side of Lake Turkana in Kenya, and Swartkrans and Sterkfontein in South Africa. Further afield, these tools have been discovered in France, Bulgaria, Russia and Spain, and in July 2018, in south China—the oldest yet found outside Africa. In November 2018, 2.4 million-year-old Oldowan blades were found in Ain Boucherit in Algeria, giving a much earlier range of tool use around Africa, and maybe the first suggestion that this tool set might have had multiple origin sites. The timescale and geographical spread over which this technology was used is huge, covering maybe more than a million years.

An Oldowan chopper

In our reading of the history of technology in humans, Oldowan tools in time are replaced with a new set consisting of more complicated kit. Thousands of miles from east Africa, St. Acheul is a suburb of the northern French city of Amiens where, in 1859, a major haul of ax heads would come to define the most common industry in the whole of human history. They weren't the first of these discovered—in the late eighteenth century, similar examples were found in a Suffolk village near to the pleasant market town of Diss—but they are the type of specimens of what is now known as the Acheulian tool set.

Acheulian hand axes have been worked more precisely than their Oldowan ancestors. Typically, they are teardrop-shaped, chiseled into sharp points, and crafted into flattened blades, often on both sides. They are also larger, with a roughly 20cm cutting edge, compared to just 5cm in a typical Oldowan blade. They represent the fruit of a concerted cognitive ability to truly craft a tool, or a weapon, and require skilled hand–eye coordination and an even greater degree of foresight and planning. The flaking of a stone occurs in multiple stages, as the initial shape is crafted, and then the blade thinned and sharpened with a second round of delicate lithic reduction. Try it next time you're on a rocky beach, with a flint. It's a difficult, skilled process; an indelicate or badly placed blow will lethally crack the stone, and maybe your fingers.

We see an increase in symmetry in these blades as brains get bigger over evolutionary time. The instruments are found distributed around the world, and across species. The oldest Acheulian tools, as of 2015, have in fact been found in Olduvai Gorge, home (at least in name) of the technology it replaced, but they are also found all over Europe and Asia. *Homo erectus* chiseled these blades, as did other early humans such as *Homo ergaster*, and Neanderthals and the first *Homo sapiens*. They used them to hunt, to butcher animals, to strip meat from skins and bones, and to carve those bones. They were used as spearheads, and some researchers have suggested that sometimes they were not used as originally intended but were ceremonial or even traded as currency.

Acheulian tools are the dominant form of technology in human history. Though there are minor refinements over time, it is truly fascinating how stable these blades are. Many more people today use telephones or drive cars, have reading glasses or use cups, but in terms of longevity, Acheulian tools win hands down. We define this period by the technology. The Paleolithic period ranges from

2.6 million years ago until 10,000 years ago. Paleolithic means "old stone," which might be slightly ironic, because much of what was being engineered by those worked stones was probably wood and bone.

So, a few decades ago, the genus *Homo* was defined by tools. But we now know that earlier apes who we don't call humans were also using stone tools. Therefore, we have to conclude that, historically, tool use has not been limited to humans. This is borne out by examples of tool use in living nonhuman animals, as we will see later. For these animals, the technological material is frequently not stone, but harvested from trees, and there is no reason to suppose that early humans were not tooling wood as well. Of course, wood biodegrades, and we have scant physical remains of prehistoric tooled wood. There is a beautiful site in Tuscany in northern Italy that has revealed some of the best examples of ancient carpentry. They're boxwood fragments, around 170,000 years old, scattered in the ground alongside Acheulian stones and bones from an extinct straight-tusked elephant, *Palaeoloxodon antiquus*. A couple of spears have been found in other sites, including in the seaside town of Clacton in Essex, but these Tuscan remains are probably multipurpose sticks, and show evidence of having been tooled, partially using fire. Boxwood is hard and stiff, and the sticks show evidence of having had their bark worn away with a stone scraper, and possibly charred to remove extra fibers or knots. Who carved these spears and digging sticks? The time and place squarely put this carpentry in the hands of Neanderthals.

These wooden tools are few and far between, especially at that age. So when it comes to naming conventions, we work with the evidence available, and what follows the age of old stone is the middle age of stone, a 5,000-year period we call the Mesolithic, and then the Neolithic and into the present.

The Paleolithic covers both the Oldowan and Acheulian tool sets, and combined, these periods represent more than 95 percent of the history of human technology. There's a measurable shift between the two types, but otherwise the toolbox of humans changes very little for two periods of more than a million years each. There are no great leaps forward in development. Humans migrated around the world during this period, reaching all the way to Indonesia, and all over Europe and Asia. We see them slowly change in their anatomy, in species and in their global distribution, but the technology remains recognizable.

With the Lomekwi tools dated at 3.3 million years old, it's worth noting that these first technological people were already maybe four million years distant from the separation of our evolutionary branches and that of chimps, bonobos and other great apes. All of them also use tools today, which we will come to in a few pages. What we are unsure of is the continuity of cultural tool use. Humans accumulate knowledge and skills and transmit them through time, mostly without losing those acquired abilities. Generally, we don't have to invent the same technology over and over again. Have all the great apes used tools continuously since that divergence, or has tool use been forgotten and reinvented many times? This is unclear, and possibly unknowable, as there is little evidence of other great apes crafting stones, even if they did use wooden tools, which are not preserved well in the fossil record. With the advent of the basic Oldowan technology in ancestors that pre-date humans, but came after the split between those apes that would evolve to become us and those that would become gorillas, chimps and orangutans, we are witnessing an ability to deliberately manipulate external objects for specific purpose that exceeds any other animal—including all the other great apes—by a significant margin.

What It Takes to Be a Maker

The scale of difference between us and the other great apes is important if our technological skills are to be considered as part of a great leap forward for us. Crafting a tool requires foresight and imagination, which needs to be translated into fine motor control actions. That is a lot of brain power to contemplate. But we must also consider the dexterity that is being enabled. When thinking about technology, we have to talk about the physicality of both brains and bodies. Our hands are awesomely complex. Roboticists try to model the number of degrees of freedom that a normal human hand has, upwards of twenty, maybe thirty, in trying to emulate what you can do without much thought. Consider the bewitching precision in dexterity that Kyung Wha Chung displays when she plays the Bruch Violin Concerto. Or when Shane Warne would spin a cricket ball so outrageously that it would turn almost ninety degrees upon hitting the ground, and utterly hoodwink the best batsmen in the world. Conjuring such magic into the muscles in our fingers and thumbs, hands and wrists requires a great deal of neurological processing, not just in the motor control, but with intention too.

We have unusually large brains. They are also unusually folded and crenellated, meaning that the density of connections between cells is extremely high, and increases the surface area of our cerebral cortex, with which modern behavior is most commonly associated. There are many metrics that can be applied to brains, and we come out near—but not at—the top of most.

We don't have the biggest brains as, in general, they increase in size as bodies do. Blue whales are probably the largest animal to have ever existed, but sperm whales have the heaviest brains, weighing in at around a monstrous eighteen pounds. On land, the heavyweight brain champion is the African elephant. In terms of the absolute number of neurons, African elephants come out on top too, with an absurd 250 billion, around three times more than ours, in second place with around 86 billion. For comparison, the nematode *Caenorhabditis elegans* is beloved of biologists for many reasons, one of which is that we have mapped the pathway of every single cell in its body as it matures from a single fertilized egg to a fully grown worm. Its whole nervous system is made up of precisely 302 cells. Don't let this apparent paucity fuel complacency: they have around the same number of genes as us, but outweigh us, outnumber us and, in terms of evolutionary longevity, will outlast us by hundreds of millions of years.

The cerebral cortex in mammals is of particular interest because of its place as the seat of thought and complex behaviors, but we're second on that chart too, this time to the long-finned pilot whale; they have more than twice the number of cells in their cortex. At this scale, African elephants have dropped down below all the great apes, four species of whale, a seal and a porpoise.

We try to compare like with like in these sorts of scientific parlor games. After all, women are smaller on average than men, and women's brains are proportionally smaller too, though—and this cannot be stressed enough—that conveys categorically no measurable difference in cognitive capabilities or behavior. So, perhaps comparing brain-to-body mass is a more useful metric in trying to establish a neurological basis for brain power.

Aristotle thought that we were the top dogs by this measure, saying in his clearly titled book *On the Parts of Animals* that "Of all the animals, man has the brain largest in proportion to his size." Aristotle was a tremendous scientist as well as being better known as a philosopher, but he wasn't quite right about that. Again, we're close but not at the top; ants and shrews beat us hands down. It was a better scientist than Aristotle who in 1871 worked this out. Again, it was Charles Darwin in *The Descent of Man*:

> It is certain that there may be extraordinary mental activity with an extremely small absolute mass of nervous matter: thus the wonderfully diversified instincts, mental powers, and affections of ants are notorious, yet their cerebral ganglia are not so large as the quarter of a small pin's head. Under this point of view, the brain of an ant is one of the most marvelous atoms of matter in the world, perhaps more so than the brain of a man.

Only about one pound in forty of our total body mass is brain. That ratio is about the same as mice, much higher than in elephants, where it's more like 1:560. The record for the lowest brain-to-body mass ratio is held by an eel-like fish called *Acanthonus armatus*. If this ignominy wasn't enough, its colloquial name is the bony-eared ass-fish.

In the 1960s we invented a more complex method of brainpower calculation. The encephalization quotient (or EQ) effectively registers the ratio between the actual brain mass compared to its predicted mass based on the size of the creature. It allows us to rank animals with a better fit that relates to observed complexity of behaviors, and in this way, we hope to get a better handle on the amount of brain involved in cognitive tasks—brains don't scale perfectly with body size or behavioral complexity. The method only really works for mammals, and lo and behold, humans come out on top. Different types of dolphin are next, then orcas, chimps and macaques.

The trouble is that bigger brains don't necessarily mean more brain cells. Density of cells is one aspect of the physiology of cognition, but there are myriad types of cells in our heads, and they're all important. It is often said that we only use 10 percent of our brains at any one time (and therefore the implication being "imagine what we could achieve if we used the whole lot!").* Alas, that is a tremendous nonsense, an appealing urban legend. All parts of our brains are used, though not all with the same ferocity at all times. There isn't a great chunk of unused hard drive sitting there lazily awaiting stimulation. The complexities of thought and action are predicated on having multiple cell types functionally connected in ways that we don't yet understand, and cellular density is not the only or defining factor in determining cognitive processing power. A study in 2007 undermined the sensitivity of EQ as well, showing that if you leave humans out of the picture, absolute brain size was a better predictor of cognitive ability, and the relative size of the cortex made little difference.

Just as with so many areas of biology, there is no simple answer to the question of how brains, tools and intelligence are related. We're dealing with some of the trickiest research areas here: neuroscience is a relatively new field, at least when it comes to gaining a precise understanding of what and how specific brain cells relate to thought or deed; behavioral psychology and ethology are difficult sciences because experiments are hard to do—there are

---

* "Imagine if we could access 100 percent" is a surfeit-of-gravitas line spoken by a typically august Morgan Freeman in the 2014 film *Lucy*. Scarlett Johansson is the titular protagonist who pharmacologically gains access to the other 90 percent, and acquires telepathy, telekinesis, the ability to somehow encounter her Australopithecine namesake, and even witness the Big Bang. It's dumb-as-bricks scientifically illiterate hooey, and highly recommended for that precise reason.

ethical constraints to consider when experimenting on people—and observations in nature are inherently limited. Brain size, density, size relative to body mass, number of neurons—all of these factors are important, and none of them appears to be the mythical one thing that sets us apart as intellectual maestros. If it sounds like I'm being cynical about these metrics, it's more that I am critical of an overreliance on any of them as a smoking gun. Big brains are clearly crucial to behavioral sophistication. But it's not all down to brains, whichever way we measure them. Evolution occurs according to one's environmental pressures, and is not in any way a predestined pathway toward the type of complexity that we have developed. Big-brained finned pilot whales, with their densely packed neocortex, will never invent violins, because they don't have fingers.

In that sense, part of the answer to the question of how we developed such artisan tool-making skills is luck. Our environment and our evolution meant that manual dexterity and brains in which the sophistication required to make and play a fiddle (a long way down the line) were things that natural selection would favor, nurture and develop. It turns out that there are tools and technology used by dozens of animals, as we shall soon see, but to arrive at the level of dexterous sophistication that is so natural to us was our path alone. It was the co-evolution of minds, brains and hands that drove us to use sticks, knap stones, refine those flakes, and eventually, after long periods of stasis, develop our technological prowess so that we could carve statues, and musical instruments, and weapons that made resources ever-more available. Despite a few animals having similarly complex brains, none has come close to our tool skills for many millions of years.

Tooled-Up Animals

The truth is that almost all animals do not use technology at all. Tool-using animals make up less than 1 percent of species. But where the adoption of external objects to extend a creature's abilities is limited in absolute numbers, there is a diversity that ranges across taxa: tool use has been documented in nine classes of animals—sea urchins, insects, spiders, crabs, snails, octopuses, fish, birds and mammals.

By the definition above—that a tool is an external object manipulated as a purposeful extension of the user's body—it's worth thinking about how the 1 percent extend their selves with technology. Here are some of the most thrilling examples.

Food Processors

Many animals use technology to gain access to food or transform it into something more palatable. The most common action is using rocks to crack or prize food from its natural container. Various macaques eat crabs and a smorgasbord of bivalve mollusks, cracking open the hard shells with rocks. They also select rocks specific to food type. Sea otters do much the same, while floating on their backs and using their own tummies as anvils. Capuchins, chimpanzees, mandrills and other primates crack nuts with stones, and some prize the edible bits from the shells with pointy sticks.

Guinean chimps use rocks to smash and chop up the fruit of the treculia tree—which is as large as a football and just as tough.

Sticks are the most common technology used by many species, for poking, gouging, prizing, scratching, digging, dragging and probing. The doyenne of primate ethology Jane Goodall has been running a field station in Gombe Stream National Park in Tanzania for more than fifty years, and there was the first to observe a chimpanzee modifying a stick for subsequent use for food processing, in this case termite fishing. David Greybeard was his given name, and Goodall watched him stripping a sapling twig in 1960, and then dipping it into a termite mound. Perplexed, she tried it for herself, and saw that the termites cling to the stick; Mr. Greybeard was eating them. Chimps also use sticks to poke honey out of beehives, and to dislodge angry bees trying to defend their homes and grubs.

Orangutans like fish and appear to like to fish. Sometimes they scavenge them dead from river banks, but they have been seen poking sticks at fish in the shallows of rivers whereon they flop into their awaiting hands. They've also been seen trying, but so far failing as far as we have observed, to stab fish in ponds with tooth-sharpened sticks; this may be a behavior they have seen and copied from humans. If this is true, it is an example of a cultural trait not only being passed between individuals, but between individuals of different species.

Plumb Lines

Orangutans and Congolese gorillas both live in dense forests, often near pools or streams that they might need to cross. Of the great apes, only humans are habitual bipeds, meaning that we alone walk exclusively on our hind limbs. The other living great apes are habitual quadrupeds, knuckle walkers that are nevertheless capable of walking on their feet, but not for very long or very comfortably. Crossing water is not easy to do on four legs as your head might be below the surface, and potentially treacherous as the floor is neither visible nor flat. Both orangutans and gorillas have been seen selecting sticks and testing the depth and lay of the floor, to determine a path through which they are wading. The gorillas may also be using them as walking sticks for support as they traverse the uneven floors of pools and streams.

General Purpose

Leaves are important as well as sticks. Orangutans are more into leafy branches it seems and have been seen using leaves as gloves when handling spiny fruit, as hats when it's raining, as cushions when sitting in spiky trees, and fashioning branches to aid masturbation. Gorillas brandish branches to ward off rivals before a fight kicks off. Chimps use layers of leaves as a kind of sponge from which they drink. Elephants carefully pull branches from trees with their trunks and use them to swat flies. The skincare regime of brown bears includes using barnacle-clad rocks for exfoliation when they are molting. In the simplest terms, these are all examples of animals using their inanimate environment to extend their own physical abilities. Whether they fashion these objects or merely use them as found, they all qualify as tools.

Sponging Dolphins

Everyone knows how smart dolphins are. They do tricks, rescue swimmers and are legendarily helpful. In all the neuroscience metrics mentioned above, cetaceans (and particularly dolphins) score very well. But despite their big brains, and complex social behaviors, sophisticated and unpleasant sexual behaviors (which we will come to shortly), and communication skills, your average dolphin merely has flippers.

Of the forty living species of dolphin, all have front flippers in which the bones are brilliantly homologous to the bones in our hands, almost perfectly like-for-like, as we both also share with horses' front legs, and bats' wings. This unequivocally shows our shared and relatively recent ancestry as mammals.* But dolphins don't have any musculature that allows differential dexterity, and flippers are fused as a flat paddle, even though the equivalent finger bones sit within. They don't do much more than flap backward and forwards to propel the owner in water. Admittedly, incredibly skillful though they are at swimming, there are few examples of tool use that don't require clamping hold of an external object in order

---

* Cetacean evolution is one of the most exciting and comprehensive evolutionary trajectories we know of. The animals that would become the whales, dolphins and other aquatic mammals branched away from the ones that would become even-toed ungulates around fifty million years ago. This means that the landlubber which is most closely related to whales is in fact the hippo.

to manipulate it. Because of their flippers, dolphins, whales, porpoises and other cetaceans aren't very good at that.

This, again, is a reminder that big brains are necessary but not sufficient to propel a species toward technological prowess. We have our hands and brains, and chimpanzees use their hands, teeth and lips to fashion sticks. Cetaceans have minimal muscular control of their jaws, and no hands. The only real example so far of tool use in these highly intelligent, large-brained mammals comes from Australia, but it is impressive and important nonetheless.

Bottlenose dolphins there do something unusual: they exploit another animal as a tool. Sponges are basal metazoans, meaning that within the animal kingdom, they are among the least sophisticated, and indeed they have no nervous system, and no brain cells at all. Bottlenose dolphins in Shark Bay hurdle sea sponges onto their beaks. Around three-fifths of dolphins are spongers in this area, and researchers think that they are doing this to protect their beaks—more technically, the rostrum—while foraging for sea urchins, crabs and other spiky bottom-dwellers that hide in the craggy seabed. They specifically select cone-shaped sponges too, which presumably sit more comfortably and securely on their beaks. One animal uses a second to eat a third.

A sponging dolphin

The spongers therefore have a very different diet to non-spongers, even within the same pods. Both forage in the same areas, so we can rule out this difference being due to ecological factors—it is as if they're going to the same buffet but choosing different food because one is using chopsticks.

How the dolphins handle the sponge and what they eat is only a small bit of this story though. There are fascinating peculiarities to be seen in this practice, and they stem from the fact that the vast majority of spongers are female. They mate with males who are not spongers, and have offspring, the females of which become spongers.

As mentioned, here we see biological transmission, and cultural transmission via learning. Some behaviors are encoded in DNA and others are acquired, yet still built on top of a genetic and physiological frame that allows the development of that trait. The scientists who have been studying this population of dolphins since the 1980s took biopsies from spongers to see if they could spot a genetic basis for their unusual foraging tool, and found none. Sponging in dolphins doesn't appear to be specifically encoded in DNA. It is entirely learned. By sampling the DNA of the spongers, the scientists could also establish the relatedness of them all, and this revealed something interesting. Sponging seems to stem from a single female dolphin, around 180 years ago, two or three generations back. We are now referring to this tool innovator as "Sponging Eve." We can see the relatedness in this group, and we can see the passage of the sponging, but also that it is not genetically inherited. What that means is that this is cultural transmission of tool use. This is the first known case in cetaceans. Daughters learn sponging from their mothers. As sponging is a cultural adaptation, it presents a bit of an evolutionary puzzle, as spongers do not appear to reproduce at a greater rate than non-spongers, which suggests that the behavior doesn't confer any great benefit or cost. As it stands though, of all the documented cases of tool use in animals, there has been almost no other assessment of the effect that it has on reproductive fitness, which is the key idea in evolutionary biology—characteristics that increase the numbers and survival of offspring are likely to be selected. Darwin's theories were formalized in the first half of the twentieth century by applying mathematical scrutiny to observations of nature. It was no longer enough to say "the giraffe's neck is the way it is because increased length has been selected as an advantageous trait for reaching succulent leaves" (see page 108). We could scrutinize and model a potential advantage by looking at how it was passed down through the generations, and if it went forth and multiplied. As far as I am aware, there is a dearth of this standard of evolutionary testing as it relates to tool use.

Cultural transmission is a hugely important idea in our own evolution. Outside of humans, so far, it's been seen in dolphins, some birds, and some monkeys. There's an artificial distinction made between biological evolution, which tends to mean genetically encoded, and cultural evolution, which tends to mean taught or learned. Instinctual behavior is knowing that food covered in fungus is likely to be detrimental to one's health; learned behavior is recognizing that ageing blue cheese is delicious. These two facets are not independent of each other, because the learned behavior has to be built upon a biological framework that is capable of acquiring and processing that knowledge. An animal needs a big brain to receive this kind of instruction and act on it.

Cultural transmission also requires innovation, and that is truly rare. We will come to our own stellar abilities in this domain shortly.

The Birds

It is notable that almost all of the examples given so far are mammals, and most of them primates. Mammalian brains are generally larger than those of other vertebrates, and also differ by having large forebrains, crammed with structures that relate very specifically to the complex behaviors associated with mammals. As we've discussed earlier, size really isn't everything. I've dissected a lot of pig's brains in my time—a creature that is often considered intelligent and social. Their brains are relatively small, the size of a plum, but encased in inches-thick cranium. If you spend much of your life butting your head up against things, a robust skull is recommended.

A macaw's brain is about the size of a small walnut, which for a bird is quite big. Birds are a large group of animals, descended directly from theropod dinosaurs, which includes the *T. rex*, the even more fearsome *Giganotosaurus*, and the much more birdlike *Archaeopteryx* (researchers consider birds to be avian dinosaurs). We know that like the small mammals, birds massively diversified after the meteorite that landed sixty-six million years ago off the coast of what is now Mexico, and called time on the big dinosaurs. There are maybe 9,000 bird species living today, which is not quite double the number of mammal species. With a group that big, there is an enormous range of diversity: the smallest is the bee hummingbird, which weighs about the same as half a teaspoonful of sugar, and the largest the ostrich (the Madagascan elephant bird was even bigger, a towering ten feet tall and half a ton in weight, but it went extinct around 1,000 years ago, mostly because we ate them). All birds today are feathered, toothless and lay hard-shelled eggs.

When it comes to cognitive behavior, our gaze has historically focused on the beasts closest to us, and the ones we are most attracted to. That means we study primates and cetaceans and elephants more than anything else. Recently though, we have turned to the corvids and parrots, and with good reason. Crows, rooks and ravens seem to be leagues ahead of most of their bird brethren when it comes to social skills and tools (raptors seem to have their own very fiery skills, and we'll come to that below). New Caledonian crows are the kings and queens of avian technology. They are known to not only use sticks to wheedle grubs out of logs and rotting bark, but to craft these tools themselves. In lab conditions and in the wild, crows will strip a twig, typically four or five inches long, until it is straight and true, and use it to root around to find food. This is an instinctive behavior—a murder of crows raised in captivity who had never seen it in another group were observed crafting and using these sticks. We also know that hooks are better than spikes. These crows manufacture and use hooked tools to fish out fat grubs, and carry them away on the hooks. In experiments where one tool was placed out of reach but visible, they used a shorter stick to retrieve the grub-fishing stick. Using a tool to carry a second object is almost unheard of in non-human animals, as is using one tool on another (a "metatool"). That shows an astonishing level of analogical reasoning which allows them to think a few steps ahead: *I know that the long stick can be used to get food; can the short stick be used to get the long one?*

Though I mentioned above that there is currently a lack of quantitative assessment of the evolutionary benefits of tool use across the board, a study in 2018 put some useful numbers onto the

Caledonian crows. Crows with hooked tools were timed to see how quickly they could retrieve either worms or grubs from a tight hole, or spiders from a wide hole. When they used hooked sticks, they retrieved the bugs up to nine times quicker than when using straight ones. This is not a metric of reproductive success directly, but an efficiency in foraging or hunting for food is just the type of thing which has a very positive effect on mating: you can spend more time foraging, and get more food, all of which make you a healthier and more attractive potential mate.

Hooks are an important technological innovation. Ask any fisherman. Orangutans may fish with their hands, or even with simple straight spears, but a hook captures prey much better than a spike. Maybe this was how Paleolithic humans first started to hunt fish rather than just forage them. We know of rich early human culture using the fruits of the seashore from the Blombos Cave in South Africa at least 70,000 years ago, which include dozens of remains of the tick-shell sea snail carefully punctured to become beads, possibly for a necklace—probably the earliest examples of jewelry. The coast provides an abundance of edible life, and we were certainly eating sessile seafoods at that time—that is, ones that are immobile, such as mollusks that can't swim away. That's a different buffet from hunting. The very first hooks, at least that we know of, were carved from mollusks in Japan. They've been found in the island of Okinawa, carefully ground out of the flat bases of shells from a different sea snail, the trochus, around 23,000 years ago. Of the two examples we have, one is an almost perfectly preserved crescent, with a ground edge that could still slice flesh. Though these are the earliest, and they probably represent a mature technology, they are a significant discovery in charting the development of us: the fish hooks, among other things, show our successful colonization into island chains, and an ability to hunt from the bounty of the oceans, rather than merely gathering from it.

No one would imagine that the ability to carve a hook from a conical snail shell was encoded in DNA. This is a skill that has to be taught, or learned, or passed on in a subculture of the broader sweep of the lives of our ancestors. Again, cultural transmission of an idea emerges as a thing that we need in order to explore our evolution. This mode of transmission of ideas isn't limited to us—as in the case of the sponging dolphins of Shark Bay. Nor is it limited to technology.

The crow's social cognitive behavior is even more intriguing. They seem to be capable of not just recognizing human faces but being able to distinguish between people who are looking at them, or looking into the distance. It's a simple experiment: scientists merely approached a murder of crows in Seattle in 2013, either looking straight at them, or not. Like punters in a fighty Saturday night pub, the birds scattered far more quickly if they were being eyeballed. Perhaps this is a recent adaptation to living in cities, in close proximity to humans, who are not always a threat. Fleeing is a costly business—it takes time and effort, energy that could be better used foraging. Pigeons and other birds with lesser cognitive abilities than corvids will just scarper in response to proximity without adjudging the intention of the person approaching. The follow-up experiment with the crows was bizarre. The researchers approached the birds wearing one of two masks. People wearing the first mask walked on by, but those with the second trapped the birds. They were conditioning the crows to recognize one face as a threat and the other as benign. Five years later, they returned to the same locations, still occupied by the same birds, and now younger ones including their offspring. The response to both masks was the same. They appeared to have remembered the threat, and somehow, they may have passed this information on to the younger birds. If these results are correct, we are yet to understand how this transmission of knowledge may work.

Despite these skills, "birdbrain" remains an insult. The origin of this cuss is not known, but we do know it was being used in the US in the first half of the twentieth century, so pre-dates our new-found interest in corvid intellect. It might simply be because birds literally have small brains, or that they're flighty and twitchy, and chickens can operate for a while after decapitation. Either way, this affront cannot fly any more. The number of neurons required for the complex vocalizations of any songbird, or even the entertaining mimicry of cockatoos or parrots was known to be high, and this posed a conundrum given the overall size of their brains. In 2016, twenty-eight species of birds had their brains anatomized at a scale never before undertaken. A neurological basis for the cognitive abilities of birds was surprisingly straightforward: what the researchers found was that the neurons are just much more densely packed. Corvids and parrots have forebrains that are comparatively the same size as the great apes, and they are crammed with neurons at a density that in some cases is higher than primates. This result may yet account for the curious case of these smart avian dinosaurs. As for the insult birdbrain, quoth the raven, "nevermore."

Now we know that many animals do in fact use tools, the key question has mutated. When thinking about our extreme abilities in technology, so extreme that it has defined our existence from chipping stones to a laptop computer, we should think less about what the tool is, and more about how that skill was acquired. Dolphins will not acquire dexterity, nor will the raven rapping at my chamber door do it with a tool much more sophisticated than a worked stick.

Perhaps it is not the use itself that distinguishes us from them. It is more that we pass on this knowledge and these abilities to craft tools.

Fiery the Angels Fell

There is a specific tool that is worth examining in more depth, because it is paradoxically destructive. The world has burned for billions of years. Fire is a relentless force of nature, a chemical reaction that destroys all in its wake, from the bonds of molecules that fuel combustion, to the life that expires in the face of temperatures living cells will not tolerate. The vital molecules of biology contort and break apart, the water in our cells boils. Fire and life are incompatible.

However, fire is part of our environment and our ecology, and the ability to adapt to, control and use such raw power is a force that has shaped evolution. We live on a mantle atop a raging molten core that has been pushing brimstone and igneous rock into the world since before life began. Indeed, we now think that the action of lava forcing its way out of the seabed four billion years ago was not just instrumental but crucial in the formation of the rocky nurseries in which chemistry transitioned into biology, and life began.* We didn't merely adopt fire; life was born from it, and is molded by it.

---

* Our current best-fit theory for the origin of life is in so-called white smokers, hydrothermal vents driven up from the sea floor during the Hadean period some 3.9 billion years ago. These towers were (and are to this day) formed of the mineral olivine and percolated with labyrinthine pores and channels driven by the tumult of the living rock below. We call it "serpentinized," and the presence of hydrogen sulfide and other charged chemicals swirling in and out of these microscopic chambers gave rise to the first cells.

Darwin described humans' discovery of the art of making fire as "probably the greatest, excepting language." He might not be wrong, though maybe today we are not quite as dependent on fire as we were during the white heat of Victorian ingenuity when he wrote those words, and perhaps we don't see open fires or furnaces nowadays as often as he did.

Nevertheless, we are pyrophiles—fire lovers. We burn the energy of the Sun that has been trapped into the carbon of wood both living and so long dead it has become compressed into coal, and in the carcasses of animals that perished so long ago that they have been literally pressured into becoming oil. It is in the destruction of the chemical bonds of those once vital carbon molecules that fire releases its energy. This has shaped the modern world and, perversely, now threatens it, as the carbon dioxide that we continue to pump out into the atmosphere itself holds more energy than other components of air, and the greenhouse effect warms our world.

Fire is a tool that has utterly transformed our existence, not just in the industrial age, but from well before our particular type of human being had settled into the form we currently enjoy. We have good evidence that *Homo erectus*, that highly successful human who walked all over the earth from 1.9 million years ago until around 140,000 years ago, was a fire user in some capacity. The dates when they first utilized fire remain disputed. Sifting through the dirt of ancient human sites is a fiddly business, and while there is molecular evidence of burnt bones and flora as long ago as 1.5 or 1.7 million years ago (depending on where you look), these are open-air sites, and it's not clear that these are not the result of wildfires triggered by lightning strikes or local volcanoes, rather than deliberate use by early humans. Some have suggested that based on the shapes of their teeth and other morphological dimensions, *Homo erectus* was cooking food as long ago as 1.9 million years. The earliest secure date for fire in an archaeological context is probably around one million years ago in the Wonderwerk Cave in South Africa.

However and whenever the transition took place, humans moved from an opportunistic use of fire to habitual use, and eventually to obligate pyrophiles. This transition, as with all the stories in human evolution, almost certainly occurred slowly and incrementally over time—there was not one single spark, but many. Archaeologists argue over the earliest evidence of controlled use of fire. Then again, archaeologists argue about a lot of things.

By 100,000 years ago, we had it largely under control. As a source of heat and light, man's red fire is of obvious benefit, as is the ability not just to control it, but to generate fire from a spark. *The Jungle Book*'s orangutan-in-chief King Louie expresses his desire to be just like you, specifically by owning this uniquely human ability, and he is wise to sing so. The impact of fire on the development of humankind is incomparable. We expanded north with fire as a source of heat beyond the temperate and tropical zones whence we evolved. This gave us access to a whole new range of beasts both large and small to hunt, cook and feast upon and make tools and clothes and art from their bodies. As is the case today, the social significance of congregating around a hearth or a fire should not be underestimated. Social bonds are forged and consolidated around a fire, stories told, skills passed on and food prepared and shared.

We are the only animal that cooks. Energy and nutrients are sometimes held deep inside the vegetation and flesh that we consume, and digestion is the process whereby they are released. This can be chemical, and mechanical. Teeth can be for grinding, tearing and chewing, but all are for some form of maceration, the process by which food is broken down to make it more accessible to the enzymes that will chew with molecular precision. Plenty of animals use artificial mechanical means to aid digestion. Birds don't have teeth to macerate, but they do have gizzards—muscular pouches in their digestive tracts that some fill with grit which grinds up food, making it easier to chemically digest. We call these "gastroliths"— stomach stones—and this is an ancient practice. The fossilized remains of many dinosaurs from the Cretaceous and Jurassic periods have been found with smoothed stones inside their body cavities, where once the soft tissue of gizzards would have been.

We outsourced some of our digestive abilities by externalizing them. By cooking foods, we break the bonds of complex molecules, and make them easier to digest in our stomachs. Meat is tenderized by heating. Softer foods are quicker to eat too, in that we spend less time chewing a boiled cabbage than a raw one, which means we get access to the essential nutrients more efficiently. Dining is a period of vulnerability: when your face is occupied with ingesting a meal, it is less alert to danger from predation. Spending less time eating means less time being potentially eaten.

All of these things make cooking a desirable and essential part of our evolution. Some researchers have suggested that we became a pyrophilic primate by living among iterant burning ecology and adapting to the benefits it brings. Some have suggested that the origins of cooking, or at least an understanding of how heat changes food, might have begun by apes foraging in burnt landscapes. It's difficult enough to roast a turkey to perfection in a twenty-first-century oven, so it's not unreasonable to suppose that animals roasted in wildfires are most likely to be burnt or undercooked. But it may be that these first hot meals were the spark of the idea of using heat to change food for the better.

The other obvious benefit of safely standing to the side of a raging furnace is that you can be presented with an exodus of other animals fleeing from the danger. If these animals are of interest to you as food, then fire provides a free all-you-can-eat buffet. We think that South African vervet monkeys do this, and enjoy an unprecedented access to invertebrates scuttling out of the fire, and into their mouths. We also think that the monkeys know this well, and so increase their normal foraging range into wildfire regions, especially after a recent blaze. There's another set of benefits from this behavior too. Vervet monkeys stand up on their hind legs to look out for predators, and thus can see over the grasses and plants. When they are cleared by being burned to stubble, the monkeys can see further. Vervet monkeys in scorched plains spend more time feeding, and feeding their young, and less time standing erect on the lookout for something that will eat them.

Even closer relatives to us, savannah chimpanzees in Fongoli, Senegal, live among fire as part of their natural ecology too. It is hot in the grasslands anyway, but since 2010, the onset of the rainy season has become increasingly erratic. From October, fires start which encroach upon three-quarters of the chimps' thirty-five square-mile range. These most often ignite at the beginning of the rainy season, when rolling thunder and lightning meets arid bush.

Scientists have been watching these chimps for decades, and in 2017 reported on their relationship with fire. There are several things worth noting. The first is that they are untroubled by wildfires. Mostly they ignore the burning brush, but sometimes wander into and explore areas that were on fire just minutes before. They appear to navigate in burnt areas frequently, which may be the same trick the vervet monkeys are using to increase their lookout range to avoid predators. We know that in the Mara-Serengeti in Kenya, other large herbivores congregate in scorched areas at higher densities than in healthy grasslands, including zebra, warthogs, gazelles and topi. It may also be easier and quicker to traverse land flattened by plants being burned to ash.

The fact that these chimps behave in a specific predictable fashion when their world burns suggests that while they cannot control fire, they certainly can conceptualize it, and crucially predict its behavior. This is a cognitive benchmark, in which the animal is capable of rationalizing and approaching something dangerous, rather than simply taking the safest course of action, which is to flee. It's a sophisticated reaction too: the way a fire burns is a complex and capricious process, which depends on what is burning, the wind, and a host of other factors, and can change in a flash. Within seconds, fires reach temperatures incompatible with life, and can release smoke and noxious gases that are also threatening to apes.

The vervets and the savannah chimps are potential clues for us when we think about the genesis of our own relationship with fire. We look to nature today to draw comparisons and speculate that what we see now might have been similar to what happened way back when. This may be egocentric. All data is useful in some way, but there is a whiff of presumption in the notion that behaviors in our fellow apes reflect our own journey to the present.

Is this what we did? Do the chimps today mimic our own evolution 100,000 years ago, or even a million? These are hard questions to answer. Behavior is not well preserved in bones nor in the ground. We can see how bodies change in relation to changing environments, such as the subtle shifts away from an arboreal life, and infer what behavior was facilitated by those bodies. We do have better clues and tools to answer the question of how fire changed us, though the evidence is almost as fleeting as wisps of smoke. We look for charred remains buried in the dirt, or for evidence of hearths and kitchens. We also look at the morphology of ancient humans, to see if cooked food was a necessity for shaping their bodies, or at least the hard parts that remain for us to scrutinize today. We can look at body mass and feeding times to construct models of the energy required to make those bodies, and calculate that they demanded particular dietary requirements. We construct tests in our living primate cousins and see how they tally with the behavior that we are just beginning to observe in the small pockets of monkeys and chimps who encounter fire on a regular basis.

These are data points that may build up a theory, but we should be cautious. Most great apes do not live on the savannah. Most chimpanzees, bonobos, gorillas and orangutans live in dense forest environments, where blazes are only devastating, and mercifully rare. There are few formal reports on the effect of forest fires on great ape lives, but peat burning in Indonesian national parks (which is associated with expansion of palm oil farming) has only had a detrimental effect on orangutans. In 2006, hundreds were estimated to have died as a direct result of forest fires.

Savannahs did expand during our evolution in Africa, forests shrank, and our morphology inched away from being adapted to an arboreal life. Single causes are rarely persuasive arguments for how we evolved to be what we are today. Though our transition into the species *Homo sapiens* occurred in Africa, I think we are moving toward our being a kind of hybrid derived from multiple early African humans. Certainly, though the strongest evidence comes from the east of Africa, we haven't really looked that hard over the rest of that vast landmass, and the earliest known *Homo sapiens* are actually found in Moroccan hills to the east of Marrakesh. What this means is that fire is undoubtedly one of the great driving forces of human evolution, but it is not the only one. Our existence in the perpetual presence of savannah fires will have profoundly changed us. But not all of our ancestors lived on the plains of Africa.

Darwin said that of all the animals, *Homo sapiens* "alone makes use of tools or fire." There, he is quite wrong. None bar us can ignite a fire or create a spark. However, we are not alone in using fire as a tool. We have already seen that the corvids are adept tool users. Until 2017, raptors—that is, birds of prey—were not known for their tool-using abilities. Raptors is an informal and broad classification, one which includes kites, eagles, osprey, buzzards, owls and so on, and therefore doesn't necessarily relay evolutionary relatedness. Owls are closer to woodpeckers, and falcons closer to parrots than either raptor is to hawks or eagles. They are all hunters though, with curved talons and beaks, and tend to have keen eyes, some with impressive varifocal vision, honed to zoom in on a tiny mammal when soaring above.

A few raptors are pyrophiles too. Fire-foraging birds of prey work on similar principles to the vervet monkeys. Tasty critters will be flushed out of a burning bush and are easy pickings. Plenty of raptors eat carrion too, and there will be lots of roasted small mammals in among the ashes. This behavior has been noted in the scientific literature as early as 1941, all over the world, including east and west Africa, Texas, Florida, Papua New Guinea and Brazil.

But some of these raptors are even smarter. Black kites, whistling kites and brown falcons all have international ranges, and are indigenous to Australia, where they hunt and scavenge carrion, particularly in the baked northern tropical savannah. These Australian lands are hot and tinder dry, and they burn regularly. Aboriginal Australians know this well and have managed the fires with great sophistication for thousands of years. They use fire to raze particular flora and encourage the growth of edible plants and grasses that attract kangaroos and emu, both of which make good meat.

The indigenous people also know the local fauna. Over a number of years, culminating in a study published in 2017, Aboriginal rangers and subsequently Australian scientists reported that black kites, whistling kites and brown falcons have all been seen doing something very thoughtful. They pick up burning or smoldering sticks from bush fires and carry these torches away. Sometimes they drop them because they are too hot—but the intention is to place them in dry grassy areas and set a new blaze. Once the grass is alight, the birds perch on a nearby branch and await the frenzied evacuation of small animals from the inferno, and then they feast.

A firehawk

Aboriginal Australians have known of these firestarters for a while.* They refer to the birds as "firehawks," which feature in several religious ceremonies, and there is a sighting in one account from *I, the Aboriginal*, the 1962 autobiography of an indigenous man called Waipuldanya:

---

* This research is led by Bob Gosford, an Australian ethno-ornithologist who lives, appropriately enough, near Darwin in the Northern Territories. Gosford and his team refer to IEK—indigenous ecological knowledge—and go to great and necessary lengths to acknowledge, engage with and build on the long-standing traditions and skills of the first peoples of Australia. This is a somewhat new practice but shows clearly how much is to be gained in the sphere of understanding our world by respecting indigenous people with humility and grace.

I have seen a hawk pick up a smoldering stick in its claws and drop it in a fresh patch of dry grass half a mile away, then wait with its mates for the mad exodus of scorched and frightened rodents and reptiles. When that area was burnt out the process was repeated elsewhere. We call these fires Jarulan . . . It is possible that our forefathers learned this trick from the birds.

In the diaphanous academic literature on this incredible phenomenon, there has been some historical dispute over whether this fire-starting is deliberate or not. This most recent study, which is the first formal scientific account, concludes from multiple eyewitness testimonies over many years that this fire-starting is fully intentional.

It is, as far as I am aware, the only documented account of deliberate fire-starting by an animal other than a human. These birds are using fire as a tool. By any of the definitions mentioned earlier of what constitutes a tool, this behavior satisfies all of them. It also goes some way to explain how fire can apparently jump over human-crafted and natural fire barriers, such as barren paths or creeks. It is possible that Aboriginal Australians learned to start *jarulan* from the birds, and later adopted it into their management of the fires that have burned throughout Australia's history. If true, this is a beautiful example of cross-species cultural transmission. It is also possible that our ancient ancestors did the same more than a million years ago, when we began a relationship with fire that will never be extinguished. Or maybe it is just a good trick, and only us and the raptors have worked it out. Either way, the ability to start a new blaze is one of the first steps in being able to control fire.

This does not mean the next steps will follow. It does not mean that these hawks are en route to forging metal or cooking food. This knowledge is one step beyond what the vervet monkeys and the Fongoli chimps do. It requires a cognitive understanding of the behavior of fire, not least how dangerous it is. But it also demonstrates an ability to plan ahead, to calculate a considerable risk. At what age would you let a child handle a burning stick? The falcons and kites are using a lethal force of nature to manipulate the environment for a meal that otherwise would have remained safely hidden in the bushes.

Fire is part of nature. The world has burned since before there was life, and nature, with its tenacious ability to adapt to the environment in front of it, has repeatedly embraced the inferno. We have gone a few steps further and created a total dependence on this raw power. There are serious health risks associated with eating only raw food. We do have other sources of energy nowadays, but we remain utterly dependent on burning the remains of long-dead animals and plants, at least for the foreseeable future. Using fire is part of our nature, and you can't start a fire without a spark. We are the only ones who can do that, but now we know we are not the only ones who see fire as a means for getting what we want.

War for the Planet of the Apes

Violence is inherent to nature. Animals clash when competing for resources, for access to females, and when hunting. The use of technology to extend the physical abilities of an animal includes weapons, which are a violent subset of tools. The delivery of a lethal blow by using an object harder or sharper than your body makes battles shorter and more effective, and therefore attractive. Among the animals that use tools, a few have adopted them as weapons. Darwin noted in *The Descent of Man* that the gelada baboons sometimes rolled rocks down hills when attacking another species of baboon, *Papio hamadryas*. Elephants and gorillas throw stones as weapons, mostly at humans it seems (they may well throw them at other uninvited species in their domains, but obviously these are the attacks we have observed rather too close-up). These are not fashioned into weapons in any particular way, but are objects that still need to be selected as potential projectiles.

Lybia leptochelis, the boxer crab, picks up and carries a pair of stinging anemones with its claws to ward off enemies, though this has also earned them the slightly less hardcore nickname of "pom-pom crabs." They'll fight other crabs if they're short of these gauntlets, and if they only have one, will rip it in half, and the anemone will grow into a cloned pair.

The Senegalese Fongoli chimpanzees who patrol among the fires also hunt with weapons that they have crafted, which is rare even within the 1 percent of animals that use tools in any capacity. Once they've identified a nest of sleeping bushbabies, they find an appropriate stick, strip it, and sharpen it to a point with their teeth. They fashion it into a spike—these spears are on average two feet long. Bushbabies are nocturnal, and ripping open a tree cavity where they are sleeping peacefully predictably results in them scarpering. So, the specific action for the chimp is to surprise them by quickly and repeatedly thrusting the weapon into the cavity with a downward stabbing action. The rapidity of the skewering denies them the opportunity to escape. The chimps kebab the bushbabies, and eat them off the bone. This, so far, is the only example of a vertebrate other than ourselves making a tool to hunt another vertebrate.

When it comes to expressions of violence, humans excel. We hunt more effectively, primarily because we have crafted evermore specific tools for killing, from the simplest clubs, to spears with Acheulian heads, to bows and arrows and boomerangs, and all the way to guns, missiles, bombs and increasingly efficient ways of slaying other animals.

In our prehistoric past we made better tools and more potent weapons. With our advances in weapons technology, we also enabled more effective ways to increase the scale of conflict. As social organisms, we organize into groups, and those groups compete for resources. Inevitably, in that competition, we turned weapons upon each other, and devised effective ways to kill our brethren. At some point in our history, intraspecific violence escalated in scale. The oldest evidence of group-level conflict—a kind of precursor to war—comes from Naturuk in Kenya. In 2012, researchers uncovered twenty-seven bodies that had lain undisturbed for around 10,000 years. When they died they were thrown into a lagoon, which has long since dried up. It was a massacre. The remains of eight men and eight women were found, with another five adults whose gender could not be determined. There were also six children. One woman was in the late stages of pregnancy, and appears to have bound hands, along with three others. At least ten of the bodies show clear signs of having met their end after severe blunt trauma to the head—the skulls are replete with fractures and broken cheekbones. Weapons of this nature would not be part of the normal hunting kit of the nomadic people that we think occupied east Africa at this time. This suggests a premeditated attack, shocking in its mercilessness, though we can never know the motivation for this slaughter.

Naturuk is the earliest evidence of a premeditated assault on a group of people. It comes millennia before we began recording our history in writing, but we can reasonably assume that group-level conflict is part of the human condition. We have been at war for the whole of history.

People have studied the causes of conflict for almost as long. All wars are different, and all are the same. Each battle is unique as a result of the participants, the technology available to them, the geography of conflict, and other factors. But the reasons for conflict are fundamentally similar. One of the earliest works of historical scholarship is *The History of the Peloponnesian War*—an account of more than two decades of fighting between Sparta and Athens, written by the great Greek historian (and Athenian general) Thucydides in 431 BCE. In it, he says that the motivations for us to go to war are fear, honor and interest.

These three are all interpretations of evolutionary themes: fear of predation in order to simply survive to reproduce and rear the bearers of your genes; honor, pride or a sense of in-group protectionism to preserve the genes that related members carry; and interest in protection of resources that enable the survival of your genes, including territory, food and, for males, access to females. However, I am absolutely unwilling to offer these very sound evolutionary theories as any sort of moral justification for human warlike behavior. Though they superficially look like the same bases for going to war, it is intellectual folly and nonsense reductionism to attribute evolutionary principles to the extremely complex political and religious reasons that actual wars happen. Nationalism is not a reasonable representation of kin selection, the idea that evolution will work toward promoting a common purpose in striving for survival in a population because the members are closely related, and therefore have a high degree of shared genes.

Countries are not kin. Humans are too closely related overall for the arbitrary, impermanent and fluid boundaries of nationhood to evoke any meaningful biological distinction on which selection could act. This is even more amplified with conflicts based on political and religious differences. Protestant, Catholic and Mormon Christians, or Sunni and Shia Muslims, are not genetically dissimilar to each other in any meaningful way. Conflicts between these groups are political, not biological, in their foundations. While there are broad genetic differences between people as you traverse the world, this natural variation has little respect for borders or beliefs, and the way we casually talk about race bears little relationship to the human variation that is present in our genes. The characteristics that we typically use in ascribing people to racial groups are visible traits such as skin color, hair texture and some anatomical characters such as the shape of the upper eyelid. These are genetically encoded, but they represent a tiny proportion of the total amount of genetic differences in humans, which are not visible, and do not conform to racial groups. The millions of people who self-identify as African American cannot be grouped together genetically in any meaningful or informative way, even though they may have darker skin on average than Americans of predominantly European descent. Most genetic variation occurs within a population, and not between them, so while it might seem trivial to label a billion Chinese people as east Asian, they are biologically a very diverse group, even though their eye shapes are more similar to each other than to many other people on Earth. Bearing that in mind, we cannot ascribe direct evolutionary motivations for war because that would require a kind of genetic essentialism or purity that is not reflected in reality.*

The death of others to help secure the survival of an organism's genes into the future is inherent to evolution. Fighting, feeding, reproduction, competition and parasitism are all primary drivers of evolutionary change. Although we see the adoption of tools for threats or actual violence, what we do not see in nature is strategic, premeditated, prolonged, armed conflict between groups of animals, which fits the definition of war.

With one notable exception: chimpanzees. While the bonobos embrace enthusiastic sexual contact to relieve tensions and conflict (examined in detail later), their closest cousins the chimpanzees are much more systematically violent. We've known that for decades, but escalation in our understanding of the degree of violence that is embedded in chimp society began just after the summer of love had waned. Appropriately enough, the contrast between the two species in the genus *Pan* has been encapsulated in the hippy counterculture phrase "make love not war," bonobos being the lovers, and chimps the warriors. It was Jane Goodall who first noted the scale of chimpanzee conflict, in the Gombe Stream National Park in Tanzania. In the early 1970s, factions in a previously united society were being seen, with a north–south rift. No one knows why this schism occurred, but it did coincide with the death of an alpha male Goodall called Leakey, who was replaced by Humphrey. Some chimps began following Humphrey, but others from the

---

* The interrelatedness of humans, race and the genetics that shape human history are explored in far greater depth in my previous book, *A Brief History of Everyone Who Ever Lived.*

south apparently perceived him as weak, exemplified by the switching of allegiance to two brothers called Hugh and Charlie. What followed were strategic raids by each side into the other's territory, targeted killings or severe beatings of male enemies and an escalation of violence into running battles. Humphrey's legion was eventually victorious, and after four years of persistent conflict, the rebels were all eliminated.

The Ngogo chimpanzees live in the Kibale National Park in Uganda. Over a decade, researchers have watched them and seen more concerted and systematic violence and apparent battle strategy. Every few weeks, young males would congregate at the edge of their territory, and in a silent single file would patrol their boundaries. During eighteen such excursions they were seen infiltrating neighboring territory and beating a male from another troop to death, tearing him limb from limb and pouncing victorious on his dismembered corpse. After ten years of these vicious skirmishes, the Ngogo chimps had fully annexed the territories that they had been raiding.

In the Mahale Mountains in western Tanzania, one group of chimps is known to have similarly encroached upon and then annexed a neighboring troop, after which all the adult males simply vanished. Like a mafia hit, no assaults were actually witnessed or bodies ever found. But it is assumed that they were killed in territorial attacks.*

The data is sparse, but we have multiple incidences of sustained lethal aggression with what scientists sometimes call "coalition

---

* In another example of chimpanzee violence, in 2017 researchers witnessed a gruesome instance of infanticide in the successful Mahale troop. Seconds after being born, a baby was snatched by a male, who was seen eating it in a tree a couple of hours later. Perhaps only five chimpanzee births have ever been witnessed by human observers, as females tend to go into hiding after giving birth, a kind of chimp maternity leave. It may be precisely to avoid this type of infanticide.

violence" (to avoid a description that invokes the very human behavior of war). There has been the suggestion that it is humans who have forced this level of warlike behavior upon the chimpanzees. By perpetually encroaching upon their territories, razing forests to the ground, by introducing diseases, by hunting them, we have driven conflict for resources in chimp troops, and killing is an incidental by-product of the escalated violence that ensues as a result. In the case of the Gombe chimps, humans had over the years handed out bananas to encourage them into areas where they could be observed.

The idea that our behavior has influenced theirs is a testable hypothesis, and in 2014 it was subjected to that scientific benchmark. If human activity were a driver of increased levels of aggression and violence, we would expect to see more of it where humans are close by. It was a hell of a study: eighteen chimp sites, analyzing every recorded act of violence and killing over a combined total of 426 years of research. They found strong links between violence and competition for territory or resources, and population density (particularly of males), and little link to proximity to human activity. In Tanzania and Uganda, both spells of coalition violence (presumed in the former case) resulted in major territorial gains. From an evolutionary point of view, this means more fruiting trees, which means an abundance of food, which means a healthier population and more baby chimps.

Therefore, lethal aggression in chimps, including coalition violence, is best understood as an adaptive strategy. In time and genes, we are close to chimps and bonobos, and the temptation to suggest an evolutionary relationship between us all to explicate on complex behaviors is ever-present. Did the predisposition toward violence exist in a common ancestor of these three groups of apes, and only bonobos have grown out of it? Or is it the other way around—was sexual conflict resolution the norm, and only bonobos have retained it? Though these are valid questions to ask, there is little data either way, and comparisons must be made with scientific caution. Let us not forget that during the six million years in which our lineage diverged away from the other great apes, they were evolving too; in the case of chimps, evolving toward utilizing violence to maximize their own survival. Their propensity for violence is a behavior that needs to be understood on its own terms, not merely as a model for understanding ourselves. We have had enough wars for the behavior of chimps to be of limited relevance to our own.

This tour through some of the less admirable characteristics of humans and other animals shows that violence, extreme and lethal in some cases, is part of the struggle for life, and it is universal. Survival is at the expense of others who do not share your genes. We often talk of arms races in evolutionary theory, as the prey will evolve to beat the assets of the predator, and the predator will evolve in turn. This eternal conflict exists within species between the sexes, and between species at every scale. Here's a cute macroscopic example: moths are prey by the dozen for echolocating bats. Arizonan tiger moths have evolved a cunning twofold ruse to avoid being eaten: they secrete an obnoxious chemical which the bats don't like, but they also emit a high-pitched sonar click that the bats can detect. Once the bat has eaten one, and associated it with the warning sound, they avoid those moths in the future. At the microscopic level, your entire immune system is nothing but offensive and defensive strategies to battle the relentless attacks by organisms that wish to continue their existence at the expense of yours. After all, natural causes of death in humans far outstrip our own determined attempts to wipe ourselves out. It is the smallest things in the living world that have had the biggest negative impact on the lives of humans: plague, Spanish flu, tuberculosis, HIV/AIDS, smallpox, and malaria—probably the single most lethal agent in our history.

Nevertheless, we have had a pretty good stab at destroying each other. What is beyond doubt is that with our brains, ingenuity and skills, we have made the act of killing more and more efficient, both interpersonally and now globally. Maybe the days of mutually assured destruction by nuclear weapons are behind us, and it doesn't take any evolutionary theorist to recognize that that is a good thing for our genes and for our species. Our reasons for going to war are difficult to justify by evolutionary theory, and this is reinforced by the fact that only chimpanzees seem to emulate a scale of conflict that could be described as anything like warfare. Most cultures agree that killing other people is forbidden, and it is even enshrined in the Abrahamic commandments, though it seems this is interpreted as more of a guideline than a rule, given the enthusiasm with which the disciples of Christ and Muhammad have engaged with the snuffing out of other people's light.

Farming and Fashion

We excel at the use of tools to extend our reach beyond the limitations of our physical forms. These abilities are almost all taught rather than inherent, but are built upon biological foundations that allow these skills to develop. As we have seen of the animals that use technology, some skills are learned, some are biologically encoded. But none comes close in sophistication. There are a couple of other characteristics worth examining that are truly part of our culture, and that may have apparent equivalents in other animals. Neither is a tool as such, but both are examples of humans extending their abilities by profoundly manipulating their environment. Both require tool use, and both are fundamentally important to humankind.

The first is agriculture. We have seen examples of organisms exploiting inanimate objects, and in the case of the sponging dolphins, one animal using a second to hunt for a third. There is another technique that we use to feed ourselves, in which we cultivate other organisms to harvest a food product. In humans, we call this farming. Agriculture changed humankind irreversibly and set the foundations for the current era. Over a short period of time, we went from being hunter-gatherers to being farmers who nurtured our own food and, in doing so, set in motion the wheels from which civilization would emerge. Agriculture has been the dominant industry and technology for around 10,000 years. When it emerged, we see the evidence of new cereal crops being bred, rye in Mesopotamia, einkorn in the Levant. We see the domestication of wild boars and sheep in multiple locations around Europe and Asia. Within a period of 1,000 years or so after the retreat of the last Ice Age, the origins of farming spring up wherever there are humans. No longer would people need to track seasons or migrating animals to secure food. Permanent bases could be established, and crops stored for fallow years. Farming requires planning and foresight, to anticipate what will grow, how and when. This in itself drives technological innovation; pots for storage, colanders for processing food, ploughs and spades for tilling the earth. The overall effect is the centralization of a valuable commodity, and that draws in more people. Economic disparity is created, and trade follows close behind. Because it was more stable, this changing lifestyle became dominant over foraging pursuits, and the practices were handed down and taught in and among families growing into communities.

Farming changed our bones and genes too. Our genomes reflect our changing diets more swiftly than many traits, and we can see the switch in agricultural lifestyles within our DNA, the classic example being in milk drinking. In Europe and in recent European emigrants, we drink milk throughout our lives. For most people on Earth today and historically, drinking milk past weaning is the source of all sorts of tummy troubles, as the enzyme required to break down a particular milk sugar called lactose works only during infancy. But at some point, around 7,000 years ago, probably in northwest Europe, people developed a mutation in the gene for that enzyme which meant that its function persisted throughout their lives. We had been husbanding dairy animals before this, and probably eating soft cheeses from their milks (processing milk into cheese removes the lactose sugar, so cheese can be consumed by anyone without effect), but not drinking their raw milk. After this mutation, combined with our farming practices, we had a new source of protein and fat, one for which we controlled the means of production. Its advantage to us is obvious, and it was selected not solely by nature but by the combination of our lives and the organisms we had crafted together. Now it is scored into our DNA.

I mentioned in a footnote on page 3 that no organism has ever existed independently of others (while challenging the notion that a virus is not classified as being alive). That is certainly true, just as predators rely on prey, and ecosystem food webs are delicately balanced networks of interdependence. Agriculture is different. It is the industrial process of symbiotic cultivation, in the sense of being systematic worked labor in order to generate a grown product. The goats we milked 7,000 years ago were being shaped by domestication, and now they are what we made them.

Agriculture has been an essential cultural development that propelled us through history and into civilization. We are not the only farmers though.

Leaf-cutter ants are famed for trotting along in television documentaries carrying colossal sections of foliage that they have snipped from a plant. However, the leaves are not the food they seek; what they want is a product made inside the cells of fungi from the family Lepiotaceae that they have crafted too, not directly, but by mutually beneficial evolution—the ants nurture the fungus, the fungus feeds the ants. Just like tilled soil, the leaves act as a substrate on which the fungus grows, which provides the ant colony's essential food.

There are around 200 species of leaf-cutter ants that do this, and it's been part of their existence for more than twenty million years. They are obligate fungal cultivars, meaning they fully depend on this activity, just as we do on farmed food. The dependence is mutual too: the fungus grows filaments called gongylidia, which are packed with nutritious carbohydrates and lipids, so that the ants can harvest them more easily to feed to the queens and larvae. Gongylidia don't exist outside of fungal-ant agriculture.

There's a further outrageous layer to this symbiosis. The leaf beds are prone to infection by another fungus, which the ants weed manually (actually, with their mandibles). But they also carry *Pseudonocardia* bacteria on their bodies and in specialized endocrine glands. These bacteria produce an antibiotic which attacks the fungal infections. This is an astonishing description of mutualism on many levels: an animal farming a fungus, using bacteria as a pesticide, each dependent on the others. Evolution is terribly clever, and we have a lot to learn from ants.

The second cultural pillar that is essential to humankind is much more elusive in the rest of nature, and it is how we choose to decorate ourselves. To dismiss how we dress or wear our hair as trivial and insignificant or of no value is daft. The often preposterous high couture of catwalks may be baffling to most of us, but physical appearance is of paramount importance to signal many messages to others. Sexual selection is a huge driver of evolutionary change, and is explored in much greater depth in the next section. As a starter though, signals that convey health, or strength, or good genes, or fecundity, allow females (generally) to select those with whom they are willing to mate. Female investment in their eggs is much greater than in males' sperm—eggs are larger and rarer than sperm, and so are a more valuable asset. This imbalance drives behavior across the whole domain of animals. One of the most visually striking manifestations of this is in the exaggerated traits that many males display. The tail of the peacock is the one we refer to most often: it's metabolically expensive to make such an ostentatious fan, and it makes it much more difficult to run away from a hungry fox when being such a brazen cock of the walk. But to survive as such a show-off may mean that you generally have good genes, and a female might be wont to think that these are worth harvesting for the best chance of survival of her own genes.*

And so, we see outlandish tails and flamboyant appendages on birds of paradise and insects of all shapes and sizes. We see crazy displays of rutting pronghorns, or the lekking of the screaming pihas, or the hilarious bouncing of the long-tailed widow bird in African grasslands. Preening males, parading how fine they are.

To a female piha, pronghorn or peahen, this might be a good look. But these are certainly not fashions. Exaggerated glamorous traits emerge slowly over generations. A random fluctuation in size of a trait might make it slightly bigger in a male, and in females, a random fluctuation in preference for a bigger one might mean that they mate with each other. Repeat over and over again across generations and the size of the trait can run away until it becomes apparently absurd. In all cases it is the increased size of the trait in a male, matched with a preference in females for that bigger trait, that drives this expensive exaggeration.

Some creatures decorate themselves. They attach things, sometimes other creatures, from the environment for various purposes, but most commonly for defense. This is distinct from, but an extension of, tool use, and seems to be primarily an aquatic phenomenon. Hundreds of crabs in the family *Majoidea* adorn their

---

* In almost all cases, exaggerated traits are in males. Females get the most return on their large investment in eggs from mating with the best males, and males are best off mating with as many females as they can. Therefore, males compete with each other for access to females, and females get to choose. This is a cornerstone of the concept of sexual selection, one of the biggest facets of natural selection. However, biology is a science beset—or enriched—by exceptions, and there are examples where the females have glamorous ornamentation. The pipefish is a kind of unfurled seahorse, and the females become more colorful when fertile, compared to the drab male. Similarly, the female Eurasian dotterel is the brightly colored one. In both cases, the males do the majority of child-rearing. Incidentally, "dotterel" is sometimes a term of abuse meaning "old fool," though this is to do with the bird's docile and unsuspecting disposition. Probably.

carapace with all sorts of objects. It's a laborious process, and their shells have fine bristles like Velcro to help stick things down. Sometimes it's simply used as camouflage, but because it takes a while, and because often the objects are stinky plants or even stationary mollusks, we think these might be used as repellents, as predators definitely know the crab is there. Plenty of insect larvae create full-body shields, often out of their own feces, which may be repellent, protective and serve as camouflage. The assassin bug carries a backpack made of the carcasses of its prey, but this is assumed to be for camouflage rather than to strike fear into the hearts of its enemies.

Outside of these assassins and soldiers, our own fashions have little in common with these adornments. Though the way we dress may have roots in concepts of sexual selection, it also may not. Some evolutionary psychologists have attempted to explain that some fashions in how we dress might reflect mating principles, but on the whole, I don't buy it. Fashions undoubtedly convey a physical enhancement which might appear to showcase a desirable trait, such as broad shoulders, small waists or apparently bright wide eyes. In his enormously popular but scientifically questionable bestseller *The Naked Ape*, the writer Desmond Morris suggested that lipstick was an attempt to make women's facial lips resemble the engorged sexually aroused genital ones. It might be a superficially attractive argument, but upon the slightest degree of scrutiny it vanishes in the haze, for there is just no evidence for this to be true. If it were true, we would expect to see selection for lipstick wearing, and higher reproductive success in women who wear lipstick. It also doesn't account for the changes in styles and colors of lipstick, or the fact that most women haven't worn lipstick for the vast majority of human history, yet still somehow managed to give birth to a healthy cohort of progeny. It is an example of the scientific sin of a "just-so story"—speculation that sounds appealing, but cannot be tested or is devoid of evidence.

Sigmund Freud attributed the necktie as being symbolic of the penis. Then again, he thought that many things were symbolic of the penis. People over the years have suggested that the phallic status of a business tie was derived from being long and thin, the fact that it hangs down and literally points to the groin, and is worn by men who like to feel powerful. But that doesn't account for bow ties. Or neckerchiefs. Or the great majority of men who don't wear Windsor-knotted ties today, or through the vast majority of human history, yet still somehow managed to sire progeny. Ruffs were popular in western Europe for a couple of centuries, but they don't point at the groin, and the Tudors seemed to procreate OK. Nor do any of these arguments account for the fact that fashions radically differ the world over and have radically changed through history. Imagine turning up to the office with a white-painted face, breeches and a giant powdered wig. Or wearing not a tie, but a small ruff, doublet and a codpiece. That a cut of trouser is in this season and not the next is much more likely to be a facet of ephemeral membership of groups. Fashions come and go, frequently. I was a goth for a while, dressing exclusively in monotones and trying my best to be morose. But then I got into hip-hop. Oscar Wilde denounced fashions as an "ugliness so intolerable that we have to alter it every six months," while almost certainly wearing a cravat, fur collar, jaunty hat and a lily in his buttonhole that may well have been ridiculed both then and now. Tribalism is a very human characteristic, and while this is not to be discarded as irrelevant to biological evolution, tribes are transient, and perhaps a good example of a behavior with which we distance ourselves from the shackles of natural selection. In non-human animals, there are almost no examples of seemingly pointless changes in behavior that mirror fashion or fads.

Take the case of Julie. In 2007, she started a brand new trend, one that has endured. Julie was fifteen at the time, a young adult perhaps beginning to grow out of the whims of playful and capricious youth. That didn't stop her from trying something new. One day, she decided that she was going to insert a blade of stiff grass in one of her ears, and that was going to be her thing. She continued her normal daily business, with the blade of grass sticking out of her ear. Her four-year-old son Jack noticed this new look his mother was sporting, and he decided to copy it. Kathy, five years younger than Julie, spent the most time hanging out with her out of all the other confrères in this group, and she adopted ear-grass soon afterward. Val was next. Other close acquaintances followed suit, eight in total out of a gang comprising twelve members.

Julie was a chimpanzee. She died in 2012, but the trend she trailblazed has persisted within her local social group and has spread to at least two other chimp populations nearby, with whom they occasionally overlap, but don't really hang out with, in the Chimfunshi Wildlife Orphanage Trust, a sanctuary in northwest Zambia. The latest reports from the primatologists who study these chimps say that Kathy and Val still wear a single blade of stiff grass in one ear.

The fashionable Julie

We observe many social behaviors in chimps that are recognizably similar to ours, many of which are discussed elsewhere in this book. This may be the only documented account of chimpanzees adopting what is described in the scientific literature as a "non-adaptive arbitrary tradition." Or in other words: a fashion. There are a few examples of other behaviors in which chimps copy other chimps for reasons that we do not understand. Tinka, an adult male chimpanzee living in the Budungo region of Uganda, has almost total paralysis in both hands, having caught them in one of the many snares that local hunters set for bush-pigs and small deer called duikers. His hands are locked in a hooked rigor with some slight movement in the left thumb, and none in the right, and they are barely functional. Tinka also has an apparent allergy, with patches of balding, rashy skin, possibly caused by mites, and almost certainly exacerbated by his inability to hand scratch or pick the mites from his body. He can't do many of the normal and essential practices of daily chimp life, which have both biological and social functions, such as grooming. Instead, Tinka has developed his own head-scratching technique, which involves pulling a liana vine taut with his foot as it's anchored on a branch, and rubbing his head against the vine, back and forth like a saw.

This is interesting as it stands. It shows a sophisticated ability to manipulate the environment and utilize your surroundings to create a necessary tool. Then again, monkeys, bears, cats and many other mammals scratch their backs up against trees, rocks or house furniture. What is much more interesting though is that once Tinka started to do this, many of his associates did too. Fully able-bodied chimpanzees copied his style, seven in total. All were younger chimps, five were female; twenty-one incidents of liana itch scratching were filmed. Tinka was present in only one of the observed scratching bouts. We cannot therefore say that the copycats were doing it as some lickspittle nod to a senior chimp. It just caught on.

These examples are few and far between. Perhaps they are outliers, freakish oddities that are not representative of the cognitive status of these chimps. But they are real. Perhaps Tinka's method is a better way to scratch one's head. The key thing is that they don't appear to be adaptive behaviors, at least not in a direct way. It looks like these chimps are copying a style for no particular reason other than to be in with the in crowd.

With tools, with weapons, even with fashion, we have extended our abilities far beyond those of other animals. While we see some tool use, flashes of the violence that we indulge in, and the merest glimpse of aesthetic choices, the differences are stark. Our cognition and dexterity have given us the wherewithal to manufacture objects of such sophistication that we became obligate tool users, creatures who have been manipulating our environment for so long that we have been utterly dependent on technology for hundreds of thousands of years.

There is, however, an older set of quintessential practices that humans enjoy. They encompass behaviors that serve a more basic evolutionary principle, ones that we have embraced and developed to a level which far outstrips its original purpose. In the next section, I will examine if and how other animals share our remarkable enthusiasm for sex.

Sex

Imagine an alien naturalist—an extraterrestrial scientist come to our world to study life on Earth, to observe us, and our place in the grand scheme of nature. The scientist would see a world bursting with life. Vibrant cells everywhere, some organized into larger bodies, but all running off coded messages inside those cells, and all interdependent. It can see through time and sees that life has existed for eight-ninths of the planet's existence, and during that time, has been continuous, with a few blips but no breaks. And it would see that none of those cells or organisms is permanent. All produce new versions of themselves, and thus the unbroken chain of life continues.

The alien scientist takes a special interest in humans, both our biology and our behavior. It notes that humans are large (but not the largest), plentiful (but not the most abundant) and everywhere (though only very recently). We are not the most numerous, by number or as representatives within our self-styled taxonomy. The mammals—hairy creatures that produce milk to nurture their young—are a small group of organisms on Earth, with only around 6,000 types known, one-fifth of which are different styles of bat. There are a few types of primate, even fewer large apes. None of them is as numerous as *Homo sapiens*, the only remaining great ape designated "human" that strode Earth's lands for the last few million years.

There have been a few members of the genus *Homo* over the years, though no definitive number of discrete human species has ever been agreed upon. Some are new discoveries in the first few years of the twenty-first century, such as the diminutive *Homo floresiensis*, the so-called island-dwelling Hobbits of Flores in Indonesia, or *Homo naledi*, a slightly larger primitive people mysteriously found deep in a crow-black labyrinthine cave in South Africa in 2013; both coincided with versions of us in time, if not in space. Then there's the Denisovans, a people known from only a tooth and a couple of bones, and their entire genome. They haven't got a species designation, because the way we classify living things relies on anatomy, and those remains are not enough. From their DNA, we know they were distinct from us and any other humans we know of. What is clear among all this murk is that we, *Homo sapiens*, are the last surviving humans, and with no plausible prospect of us diverging into new, sexually incompatible populations, we will be the last humans.

Despite our apparent ubiquity and success, the curious scientist would see that we are not creatures with great durability, at least so far. We are a tender 300,000 years old; though our larger familial group—the great apes—has endured a much sturdier ten million years. By comparison, the dinosaurs, which we sometimes mock for not having survived an interplanetary impact the like of which has not been seen for sixty-six million years, were a class of animal whose tenure on Earth far outstrips our own; we have not had to face the consequences of a meteorite the size of Paris. In fact, the longevity of dinosaurs was such that we humans are closer in time to the mighty *Tyrannosaurus rex*, than the mighty *Tyrannosaurus rex* was to the iconic stegosaurus.*

---

* Roughly: dinosaurs spanned from around 250 million years ago (mya) until 66mya. Stegosaurs: 155–150mya; tyrannosaurs: 68–66mya.

In trying to piece together universal rules about why all of these creatures behave as they do, the alien would see a diverse range of abilities and lifestyles. Upon even the most superficial inspection, one aspect of human behavior would be utterly impossible to ignore. We spend a titanic amount of time, effort and resources on trying to touch other people's genitals.

If our extraterrestrial researcher is not a sexual being,* this is all a bit of a puzzle. They note that there are two different types of human for the most part (though throughout history in every culture there have been those who either biologically or by choice are somewhere in between). They see that a large proportion of humans don't show any particular interest in sex at all until their second decade, at which point almost all of them do. The alien likes data, and observes that once they start expressing an interest, most members of the human species have fewer than fifteen sexual partners during a lifetime.† They also note that they like touching their own genitals: almost all humans that can masturbate, do.

So, from an outsider's point of view, sex is a huge, vibrant part of the human experience. Some of the specific actions of genital touching have existed in the sea eons before anything vaguely hairy ever walked on land; in fact, before trees existed, and before the

---

* Plenty of complex organisms are not. Rotifers, for example, are tiny wormy things, a tenth of a millimeter long, and found almost everywhere there is fresh water. Hundreds of rotifer species are all females, having ditched males as unnecessary some fifty million years ago. They seem to be doing fine.

† Again, there is not a great deal of detailed data on these sorts of questions. But what we do know is quite revealing. According to the mathematician Hannah Fry, studies have put the mean number of self-reported sexual partners at around seven for heterosexual women and thirteen for heterosexual men, though she notes that some (particularly men) claim many thousands, which means that the mean is not a very useful stat in this case. We also know that women tend to report specific numbers, counting upwards, and men tend to round up, often to the nearest five. Both are valid estimation techniques, but the women's technique is prone to underestimation, and the men's is prone to overestimation. Funny that.

current continents were formed. The alien observes the huge, fearsome, armor-clad, razor-toothed *Dunkleosteus*, a Devonian fish from some 400 million years ago that copulated by ventrally inverting with its partner—that is to say, an early fishy version of the missionary position that many sharks still go in for today—so that penetration and internal fertilization can occur (like many living fish, the males also have rather sturdy "claspers" so that they can hang on).

There is an orgy of ways in which genital tactility can occur, in any combination of the two sexes in humans and in other animals, but the act of sexual penetration is very old. Nevertheless, it is one that humans continue to enjoy. The statistician David Spiegelhalter has puzzled on the numbers that describe our sexual lives, and estimates that something like 900,000,000 acts of heterosexual intercourse take place per year in Britain alone, or roughly 100,000 per hour. If we extrapolate that to the seven billion humans alive, it works out at around 166,667 every minute.

Why would this bipedal creature dedicate such industry to physical communion of this nature?

Of course, everyone knows the answer to this question: sex is for procreation. It is for every sexual species. A combination of genetic material supplied in eggs and sperm seeds the growth of new, but subtly different, versions of the same creature. The primary purpose of sex is to make babies. Females wish to have sex with males, and males wish to have sex with females. Between these two pillars of evolutionary necessity, there lies a multitude of sins.

It barely needs saying that not all sexual acts in humans occur specifically to make babies, but we do them for other obvious reasons; for fun, for bonding, for sensory stimulation. The odd thing about the frequency and effort devoted to sex in humans is that our extraterrestrial anthropologist would struggle to arrive at the conclusion that any sexual act was ever followed by pregnancy and the arrival of a small human. In Britain, around 770,000 babies are born each year, though if we include miscarriages and abortions, the number of conceptions rises to about 900,000 per year.

What that means is that of those 900,000,000 British bouts, 0.1 percent results in a conception. Out of every thousand sexual acts that *could* result in a baby, only one *actually does.* In statistics, this is classed as not very significant. We are only considering heterosexual acts of vaginal penetration here, so to include homosexual behavior, and sexual behavior that cannot result in a pregnancy, including solitary acts, then the volume of sex that we enjoy magnificently dwarfs its primary purpose. So can we truly say that sex in humans is *for* procreation?

Humans are different from other creatures. By engaging in acts that don't directly enhance our own survival, we have loosened the shackles of natural selection. Evolution of humans in the last few millennia has been a complex partnership between our more basic biology and the culture that we have shaped and crafted with our intellect, graft and ingenuity. That has meant that the drive to reproduce, to simply be husks for the propagation of our genes, has been complicated and disturbed, at least compared to what came before.

Nevertheless, no one could argue that we have not been a fecund species. There are more people alive than at any other point in history. Until 1977, all of them occurred following a man and woman having sex.* The rate of population increase has accelerated

---

* The advent of in vitro fertilization was marked by the birth in July 1978 of Louise Brown, who was conceived the previous November. This is still the fusion of egg and sperm, provided by woman and man, so remains sexual reproduction. Some estimates suggest that more than five million IVF babies have been born since. I am sometimes asked if IVF, and specifically selection of embryos free of certain diseases via the technique called "preimplantation genetic diagnosis," will have a significant effect on human evolution. I think the answer is no, because the numbers are relatively small, and only accessible to a tiny proportion of humanity, as it is a technical and expensive procedure.

alarmingly. We hit our first billion at the beginning of the Victorian era, and our second by 1927. But the gaps between our second and third, and all the way to the seven billion humans alive today, have got smaller and smaller. Most of this is to do with our brilliance at dealing with disease, infant mortality and death, rather than us having a lot more sex. The widespread use of effective contraception does not appear to have significantly dented population growth, though may yet have an impact as we globally attempt to balance available resources with our desire to have sex and procreate. Statistics on our sex lives are hard enough to come by in the twenty-first century, let alone in the past, but there is little to suggest that we are having significantly more sex than ever before.

When it comes to sex, there is a galactically skewed ratio of reproductive acts versus all other sexual activity. When thinking of our sex lives in relation to the rest of the natural world, the question becomes: "Is this normal?" We spend so much time engaging in sexual activity, and yet so little of it results in babies. Sex is a biological necessity, and our interest in sex has clearly evolved well beyond any basic animal instinct. But we are animals. Has our obsession with sex made us different?

The Birds and the Bees

Let's start with the basics of sexual reproduction, which might appear simple enough, but across the animal kingdom are in fact incredibly diverse and messy. Some of the descriptions of sex that follow will sound familiar to us, other acts less so, I hope. But to get to grips with the complexities of our own sexual behavior, we must indulge in a brief survey of the sex lives of other animals.

There are many ways to be sexual, though they broadly fit into two categories. The first of these is species that have two sexes, which we traditionally call male and female. In mammals, sex is determined by discrete packages of DNA called chromosomes. We inherit a set of twenty-three chromosomes from each parent, which are matched as pairs, except for the fact that one of the pairs is not a matching pair half of the time: females have two X chromosomes, but males have an X and a Y. The female egg contains one set of chromosomes, including an X, and every sperm contains another set, each one either bearing an X or a Y. In reptiles, birds and butterflies, it's the other way around (with slightly different but irrelevant annotation: males are WW, and females are WZ).

But that's not the only way to determine sex. In some animals, maleness and femaleness is not governed by the presence of certain chromosomes, but by where you are conceived: for lots of reptiles, sex is temperature dependent, meaning that differences of as little as one degree Celsius in where an egg is placed in relation to others will determine whether that egg is a male or female. For some reptile species, an egg in the center of a clutch will be slightly warmer, and will therefore develop as male. For the strange New Zealand reptile the tuatara, it's the other way around. In crocodiles, you become a female if you're a particularly hot or cold egg, and male if you're somewhere in the middle. And so it goes. Ours is but one of myriad ways that males and females can be made.

In the second broad category of being a sexual organism are the species that have dozens of sexes, possibly thousands. Mostly these are mushrooms and other types of fungi, which we don't normally think of as sexual but are nonetheless. They have what are called "mating types," which are sections of DNA that are variable between individuals, and simply indicate to a potential mate that they are different enough to warrant sex. Mates are hard to find if you are a mushroom, as they're pretty slow-moving, and sex doesn't happen very often, so a rare chance encounter with another lonely mushroom who happens to be the same sex type as yourself is a disaster. It therefore pays to have as many options as possible, and the best way is for you to have many mating types as long as none of them is yours.

Mushrooms aside, most sexual organisms fall into the male-and-female category. Compared to the many permutations of fungi, when it comes to sexual reproduction that includes males and females, the act itself is mesmerizingly diverse. Penis in vagina is but one way. That's an old technique, as in the case of the prehistoric *Dunkleosteus*, mentioned previously. Many insects, such as the bedbug *Cimex lectularius*, aren't that fussed about a specific entrance for penetration, and a male will pierce the abdomen of his mate with a very pointy scythe-like aedeagus (equivalent to a penis), and the sperm will find their way to the eggs via the internal organs of the female. We call this "traumatic insemination."

Plenty of animals don't engage in penetrative sex at all, and external fertilization is how it's done. As with many fish, male and female Chinook salmon release their sperm and eggs into the water, and the ovarian fluid that envelops the eggs acts as a highly selective filter; some sperm are simply able to swim faster through this gel than others, and this skill appears to be determined genetically by the females: their fluids act as a filter for the best and most genetically suited swimmers. Birds tend not to have penises, and so transfer sperm via a "genital kiss," where egg and sperm meet near the cloacal entrance/exit, and are internalized by the female. That's the case for most birds, but not all. The Argentine ruddy duck has a corkscrew penis, which twists in the opposite direction to the female's tortuous corkscrew vagina, thus allowing her to retain a degree of control over who sires her ducklings.

With competition for reproductive rights being so fierce, some aspects of sex are not designed solely to impregnate the female, but to simply prevent another male from becoming the father. As in all sporting competitions, there are defensive strategies and offensive strategies. In defense: many creatures across the whole domain of animals use copulatory plugs, physical barriers inserted after sex to prevent another male from delivering their sperm successfully. In offense: some male flies release toxic semen to poison any further attempts. Some fish and some flies store their sperm in compartments, and can regulate how much they release, which is dependent on how many males have had sex with a female already, and where they are in the rank. The simplest tactic is to merely hang around long after you are welcome, and in some cases while still locked in coitus. Dogs do this, and sometimes you might see a pair of them *in flagrante* for half an hour or so in a local park, the female dragging the male around on her back. Dog penises have a section of erectile tissue called the knot (or more properly, bulbus glandis) which helps the dog sustain an enlarged erection after ejaculation, causing a vaginal anchor for a time, which has the simple effect of preventing another male from assuming the same position. Not particularly sophisticated, but quite effective.

While there are many ways that males and females can do it, plenty of animals aren't so binary. Sexual reproduction that features two sexes doesn't necessarily mean that there are two different types of organism or gender. Many creatures are hermaphrodites, and bear both sexual types in one body. Flowering plants are like this of course, carrying both pollen and ovules, the botanical equivalent of sperm and egg. The earliest recorded example of sexual reproduction comes from an alga, with the perfect name *Bangiomorpha pubescens*, fossilized around a billion years ago in what is now Canadian shale. In microscopic slices through these fossils we can see sexual spores, equivalent to sperm and egg.

Female Komodo dragons are capable of parthenogenesis when the situation requires it. This means that they can conceive without ever having had any contact with a male—literally a virgin birth—and in the absence of a sex chromosome from a father, all her offspring will be male. Komodo lifestyles are fairly solitary, and they may not come across potential mates very often. This way, she can mate with her sons without needing to have encountered a male (though this is a last resort, as it is not a great idea over multiple generations; with no new genetic information that would be provided by a father, they soon become profoundly and dangerously inbred).

Then there is the drama enacted by organisms such as the flatworms. When a couple of hermaphrodite *Pseudobiceros hancockanus* feel the urge to reproduce, two of them coil around each other and engage in an aggressive head-to-head wrestle with weapons drawn, an act which has the scientific designation of "penis fencing." Whichever worm wins does so by piercing the head of the other with its spiked organ, and coerces the loser into adopting the female role in this relationship, and becoming the sperm receptacle and egg bearer. It's easier to produce sperm than it is eggs, and it's harder to bear the younglings, so the individual who managed to claim the male mantle remains childfree, and ready to crack on for another round with a new individual. And they say romance is dead.

On the great evolutionary tree of life, flatworms are animals almost as far away from us as any, but penis fencing occurs in much more closely related beasts, including the mammals. Plenty of whales lock horns in this manner, and our even closer mammalian cousins the bonobos touch swords to resolve conflict, to make friends, and even when getting excited about a forthcoming meal (though this jousting is merely competitive rather than resulting in full penetration).

Wrasse, groupers and clownfish are sequential hermaphrodites. These fish tend to have strict social hierarchies, with a dominant female who mothers the whole brood. If the dominant female clownfish is removed, perhaps by being eaten, then as a result of an absence of her hormones among the group, a male—typically the largest—will move up a rung on the heavily stratified social ladder, and will spontaneously undergo a radical sex change. His testes will atrophy and he'll grow ovaries, and over the course of a couple of days, he becomes she. The fish will balloon, becoming a replacement dominant female.*

Social structure in nature plays a big part in how the sexes are organized. Bees, wasps and ants have two sexes, but equality is far

---

* Removal of the dominant female clownfish is the opening to the excellent 2003 film *Finding Nemo*. The plucky titular protagonist is the smallest and only remaining offspring in the shoal and is then raised by his father, before an exciting adventure unfolds. The biologically accurate version of this film would have the father, Marlin, physically transforming into a female, and then having sex with his own son, but I guess that would be a different, possibly less popular, story.

from their hive-minds. Males only have half a genome, and their lives are defined by just two jobs: protecting the queen and the colony, and having sex with the females on demand. They are literally sex slaves. If these insects seem far from our own clade, two species of mammal also use a similar system. The social structure of naked mole rats and Damaraland mole rats includes a fertile queen and a couple of mating males, while the rest are sterile male workers—some tunnelers, others soldiers.

Being a sex slave might be a better deal than the male Australian redback spider gets, whose best evolutionary tactic is to provide the ultimate dinner date—immediately after releasing his sperm to a receptive female he is eaten by her. If she is eating, she is preoccupied, and sated with nutritious food that will help nurture her spiderlings, and therefore she is less likely to mate with another male spider, who might displace the first one's sperm. This strategy is known as "reproductive cannibalism"—possibly the least sexy phrase ever devised.

Another model of courtship in animals has a much better scientific name. In socially stratified organisms, it pays for females to occasionally mate with a male who is not an alpha, but this is not always easy—and is potentially lethal—for a sub-alpha male. Plenty of tactics exist to distract the dominant males long enough to sneak a quick sexual encounter. Barn swallows will raise the alarm call of an aerial threat from above, in order to quickly and cautiously mate, while the duped birds are evading a fictional attack. Most demonstrative of all are the cuttlefish *Sepia plangon*. Males looking for opportunistic and safe sex with a female will change their color patterns on the side facing the dominant males so that it resembles a female. The dominant males assume no mating competition, and the temporarily effete male gets potential access to a female that would otherwise result in the furious ire of a dominant male. This heinous perfidy is officially known as "kleptogamy"—stolen mating—but no one calls it that. The great evolutionary biologist John Maynard Smith gave it a much better name, which is used universally in evolutionary circles: the "sneaky fucker strategy."

You might wince—or nod appreciatively—at these acts being so apparently similar or different from the way we do it. It might seem tempting to assume that a recognition of some of these behaviors in us indicates a common ancestral route. Here we must be careful. Reproduction based on organisms having male and female parts is clearly an ancient business, but the details of how this plays out in specific organisms may be independent of each other. The sex acts we see in nature do not necessarily have homologues in us, no matter how similar they may seem.

Of course, the most numerous and successful domains of life—the bacteria and archaea—do not have sex at all, and merely undergo binary fission, splitting in two to pass their genes into the future.*
But among the beasts (and plants and fungi), sex is clearly a smart trick to have in your evolutionary backpack, and has evolved in a multitude of ways that appear sometimes familiar and sometimes totally alien to us.

---

* They do a sort of version of sex, where genes can be passed from individual cell to individual cell. One extends a pilus (Latin for "spear") and passes a small loop of DNA from one to another. This process, called "horizontal gene transfer," is why we humans currently face a great crisis with antibiotic resistance. Once a useful trait (such as resistance to an otherwise lethal drug) evolves in one cell, it can be passed around quickly and at will. This is also the reason why the root of the tree of life, before complex life arose some 2.4 billion years ago, is nothing like a tree at all, but instead a matted web, with no distinct branches, just a net of gene flow from every one of the billions of cells under the sun.

Autoeroticism

The primary reason for sex is reproduction, and while there are myriad ways to make babies in nature, in humans, as we have seen, sexual activity almost never does. The question is why there is so much sex that clearly cannot result in offspring.

Unlike flowering plants, rotifers, or Komodo dragons once in a blue moon, self-pollination is not something that we can do, and autoeroticism will not work. The numbers on masturbation are again a little fuzzy; many surveys have been done over the years, with variance in how the survey was conducted, what questions were asked, age ranges and many other variables. Almost all suggest that the majority of sexually capable people have masturbated in the previous year. Some polls have suggested that more than 90 percent of men have done so in the same period. We could filter these stats by many strata, but I'm choosing a conservative number here, from the United States National Survey of Sexual Health and Behavior, which reports that apart from three age groups,* men and women had all masturbated alone at least once in the previous year.

---

* Only for women younger than seventeen, women over the age of sixty, and men over the age of seventy did the number drop below 50 percent. Survey results vary for many reasons, including the survey being done face to face, where numbers drop, and the tendency for men to overestimate sexual activity and women to underestimate it.

One of the reasons why we struggle with the numbers is because of the associated historical stigma.* While the great ancient Greek anatomist Galen recommended masturbation for women to release bodily tension, Samuel Pepys was not quite so comfortable in documenting his own solitary behavior, and wrote it in a secret diary in code. From the early eighteenth century and for a long time afterward, masturbation was viewed by the Church in Europe as a colossal sin, and by others as terribly bad for your health. The Swiss doctor Samuel-Auguste Tissot wrote an influential treatise in 1760 on the profound dangers of onanism, which included the very specific assertion that losing one ounce of sperm is worse for your health than losing forty ounces of blood. This, I hope I don't have to point out, is very much not true.† John Harvey Kellogg, the founder of the breakfast cereal empire, was similarly concerned with the damaging emission of humankind's precious bodily fluids, and alongside creating the cornflake, he invented foods that he hoped would contest the evils of male self-pollution, and some startling anti-onanistic apparatus—a kind of metal sheath with spikes on the inside just in case the wearer got an erection.

Those who fought to end masturbation were hopelessly contesting a relentless tide. Whatever the precise numbers, it is not unreasonable to state that most humans who can masturbate, do.

---

* Statistician David Spiegelhalter's wonderful book *Sex by Numbers* (Profile, 2015) is the essential recommended reading here, and on all matters concerning human sexuality. It is rigorous, penetrating, thorough, robust and extremely amusing.

† Tissot also wrote in the book *L'Onanisme* that masturbation resulted in "a perceptible reduction of strength, of memory and even of reason; blurred vision, all the nervous disorders, all types of gout and rheumatism, weakening of the organs of generation, blood in the urine, disturbance of the appetite, headaches and a great number of other disorders." If you suffer any of these symptoms, please seek medical help.

But this solo behavior is certainly not limited to humans. Though it would be easier to simply list the animals that masturbate, because it is so common in nature, I will briefly describe some of the most enlightening cases.

Many parents will have faced the awkward moment in a zoo when having to explain—or distract a child from—a male primate masturbating in full view. Males of some eighty species, and females of around fifty species of primates are known frequent masturbators. Manual dexterity may well make it easier, but evidently hands are not essential for solo sex, and many cetaceans in captivity rub their genitals against a hard surface until they ejaculate. Male elephants have prehensile muscular penises to allow them some navigational control in the long and bent elephant vagina, and with these muscles, young males rhythmically bash them against their own bellies to completion. Male Adélie penguins in the Antarctic will gyrate and rub themselves, and spontaneously spill their seed on the ground in the absence of females.

The parsimonious explanation for the volume of masturbation in humans is that it is pleasurable. We cannot ask animals if they enjoy solo sex, and we have difficulty in assessing pleasure by any means. The question is, why do they do this at all?

For marine iguanas of the Galapagos, there's a very good reproductive reason for self-stimulation. Females only copulate once per season, and the males need pretty much exactly three minutes to ejaculate. Big males will often yank a mounted smaller one from a female's back before he's finished copulation, literally pulling them off. But the smaller male iguanas have a cunning tactic. They ejaculate before they begin sex, and store the sperm in a special pouch, so that under the time constraints of being bullied off the job by a bigger male, they can just sneak the vital package in without needing the full three minutes.

There are plenty of non-reproductive emissions though, with some interesting ideas in the academic literature to explain them: a release of extra or unwanted sperm is one; spontaneous ejaculation as a sexual display is another—an African antelope, the tsessebe (*Damaliscus lunatus*) ejaculates after smelling a female in oestrus but before mating as a kind of display. Male Cape ground squirrels are the other way around, and finish themselves off a second time after copulation with a female. These squirrels are very promiscuous, and dominant males particularly so. The best theory to explain this is that it is for hygiene reasons, a means for the males to protect themselves from sexually transmitted diseases by flushing their tubes.

If these examples all seem to fit into existing evolutionary paradigms, plenty of masturbation does not. There is a slight scientific reluctance to consider the pleasure principle—that an action simply feels nice. Perhaps this is because pleasure and all emotions occur in our minds, which are difficult to access for other creatures. Humans can express joy by saying so, and saying so precisely—*this feels good*—and we trust that that sensation is real. We can only use proxies to assess the emotional status of animals. It might feel obvious enough; a stroked purring cat or a dog's bounding enthusiasm for its human. It is not relevant that we have bred these characteristics into domestic animals over thousands of generations, or that the expression and seeking of that pleasure is simply for interspecies reciprocation. Neither of these very plausible reasons for the existence of an emotion preclude the emotion being real. Experiments that strive to scientifically assess the emotional state of an animal are few and far between; on page 199 we will find out about measuring disappointment and regret in a rodent, but there are studies that indicate that rats enjoy being tickled by humans. They make an ultrasonic peep that sounds a lot like laughter, and do spontaneous leaps called *freudensprünge*—literally jumps for joy. To my knowledge, no studies have been done to address the question: "Does masturbation make you feel good?"

I suspect that in separating ourselves from the rest of the natural world, we might not allow ourselves to entertain the notion that very human-like sensations are also behind similar behaviors in animals, at least sometimes. In science, we like to generalize and find rules that cover a multitude of observations. We push back against anthropomorphic explanations, and I'm cynical about an over-reliance on adaptation explanations that are too neat, and too Panglossian. Certainly, some autoeroticism sits easily among the copious amounts of non-reproductive sex that does have an evolutionary strategy underwriting it. But masturbation is so widespread, and at least in mammals features such creativity, that generalized adaptive explanations fall short. Emergency-room doctors will tell tales of patients presenting with unusual injuries caused by non-obvious and imaginative forms of self-stimulation. The great Alfred Kinsey, who pioneered a scientific and unashamed investigation into sexual mores from the 1950s onwards, asked questions in his sex surveys about male masturbation that involved inserting objects into the urethra. We must not judge, and credit for resourcefulness must be given to the cetacean mammals. There is one reported case of a male dolphin masturbating by wrapping an electric eel around his penis.

Mouthing Off

Sexual acts are not limited to the anatomy of the genitals. Mouths are complex anatomical structures with multiple mechanical features, jaws, lips, tongues, teeth. They are also hotbeds of sensation, richly innervated for touch, temperature and taste. These features of the oral orifice mean that mouths are useful not merely for eating, or communication, but can serve as organs of action, and that includes sexual acts. Oral sex doesn't appear to play a big part in the various annals of erotic history, nor in the art of classical Greece and Rome. Perhaps hygiene played a role in those historical preferences. Again, statistics are not easy to come by, but one recent survey of more than 4,000 men and women suggested that more than 84 percent of adults have engaged in either fellatio or cunnilingus, neither of which can result in a baby. Is the near ubiquity of oral sex limited to humans? Again, the answer is very clearly no.

To begin, let's briefly look at an act of oral sex that has anatomical limitations. Henry Havelock Ellis wrote about goats in his 1927 book *Studies in the Psychology of Sex*:

> I am informed by a gentleman who is a recognized authority
> on goats, that they sometimes take the penis into the mouth
> and produce actual orgasm, thus practicing auto-fellatio.

That's a neat trick if you can do it. Though physically challenging in humans, the Kinsey Report found that 2.7 percent of the male responders had successfully self-fellated. When I was at school, the almost certainly untrue legend was that sex-addled rock god Prince had a rib surgically removed in order to orally self-pleasure. The concept is certainly not alien to human culture, in the most fundamental way. All societies have creation myths, not all as mundane as the Christian, universe *ex nihilo* version of something from nothing, Adam sculpted from red earth. Rather more dramatically, Atum, the self-created Egyptian god, performed auto-fellatio, and spat his ejaculate out, which separated to become the gods of the air and water. People have clearly been thinking about this for a long time.

Autofellatio may be rare in humans, and autocunnilingus surely is close to physically impossible (as far as my research goes, there is no academic literature on this). Nevertheless, oral sex is a common and popular version of non-reproductive sex between human partners. We do it because it's pleasurable, but we are far from unique in performing this act. Oral sex is widespread among animals, and the reasons are more difficult to scrutinize. Heterosexual oral sex is common, strikingly in the fruit bat *Cynopterus sphinx*, where females lick the shaft of their partner's penis during penetrative intercourse (done dorsoventrally, that is, from behind, but they do it hanging upside down). This has the possibly counterintuitive effect of prolonging sex. There are plenty of scientific theories why they do this, none of which can be summarized as "because they can," disappointingly. Prolonged intercourse may increase the likelihood of fertilization, or it might be an act of mate-guarding—that is, preventing another male from having a go. It might even be another way to prevent sexually transmitted diseases: fruit bat saliva may have antibacterial and antifungal properties, so to add it to the vaginal lubrication during penetration might be a safe-sex method of guarding against chlamydia and other infections.

While not quite oral sex (for anatomical reasons), the male dunnock *Prunella modularis* often pecks at the cloaca of females to remove the sperm of a rival male. These rather dull hedge sparrows have sex up to a hundred times a day, an impressive fact that is only slightly undermined by the knowledge that each act takes about a tenth of a second.

If both of these acts of oral sex seem quite functional and untender, the first report of oral sex in bears may offer a different perspective. Published in 2014, it details how two unrelated male brown bears engaged in repeated daily bouts of fellatio for six years in Zagreb Zoo, Croatia. The giver and receiver was always the same way round, and the act itself fairly ritualized into a predictable pattern. One would approach the other who would be lying down on his side. The giver would physically part the receiver's rear legs, and begin fellatio, often while humming. It typically lasted between one and four minutes, and clearly appeared to result in ejaculation by the receiver, as qualified by muscular spasms. These bears were raised in captivity, and again, this behavior might be abnormal, at least compared to bears living in the wild, where ursine fellatio has never been observed. The researchers speculate that it might have started in an absence of maternal suckling behavior, the bears having been orphaned when cubs. However it originated, it seems perfectly plausible that the behavior continues because it is pleasurable.

Pleasure is the reason we perform oral sex. Again, as with examples of masturbation in other animals, the idea that motivation for non-reproductive sexual acts might be the same as in us is not something that scientists are drawn to. Whether this reluctance is valid or not is difficult to ascertain, but certainly, examples where parsimonious explanations can plausibly include pleasure—such as in the Croatian bears—are rare. We need to be more open to the possibility that some animal behaviors, sexual or otherwise, might be driven by pleasure, but we also need to be better at assessing it. Until then, the joy of sex is limited to us, mostly.

Whole Lotta Love

Autoeroticism, fellatio, autofellatio; a compendium of examples of non-reproductive sex could go on and on. The wild party of sexual behaviors in the natural world defies our imagination, and while marveling in this panoply is fun, the point is that sex has evolved to be much more than simply an act of reproduction for most animals, and that includes us. That is not to say that the myriad purposes of these acts are the same, nor that similar acts are rooted in the same evolutionary origin. It appears that some of them, notably the many autoerotic acts, may simply, like in us, exist because they are pleasurable. We should not make the mistake of assuming that all behaviors have some specific evolved function: animals can enjoy sensory stimulation too. Rats enjoy being tickled, cats purring, and the Croatian fellating brown bears certainly appear to be enjoying themselves.

Humans engage in a huge range of sexual behaviors, most of which are not reproductive, and some of these are seen selectively in animals. There is little dispute that a healthy sex life between people helps pair bonding and stability in relationships, which may or may not be homo- or heterosexual, monogamous or polyamorous, or other combinations I haven't thought of. So while the pleasure of sex accounts for the sheer volume, in many circumstances, a secondary function is in the reinforcement of social bonding, mostly between couples. Aside from us, only one animal engages in quite such a large sexual repertoire with comparable enthusiasm, and so the question for ethologists and psychologists is whether they do it like they do it for similar reasons. The bonobo, *Pan paniscus*, is the fifth member of the remaining great apes, along-side us, gorillas (*Gorilla gorilla*), orangutans (*Pongo pygmaeus*) and the chimpanzee (*Pan troglodytes*). In fact, bonobos resemble the chimp so much so that they used to be called pygmy chimps and were only designated a separate species in the 1950s. They are not significantly smaller than their genus cousins. They are morpho-logically different, though not by much: bonobos are exclusively arboreal, living in small groups in only one forested region by the Congo River in the Democratic Republic of Congo, where there are fewer than 10,000 remaining. They tend to be less muscular than chimps with narrower shoulders, and slightly longer arms, more gracile. They have pink-red lips and dark faces, and often sport a neat center parting on their tufty head hair.

Like all great apes, their society is highly structured. Unusually, bonobos operate within a matriarchy. Dominant females reign over social groups, and bestow male status depending on their relation-ship with the senior females. They form tight-knit groups and exert control over males, especially with regards to aggression and mat-ing requests. Also unusually, for primates, as females mature, they move away from their natal groups and set up in a new clan, if welcomed by the ruling matriarchs.

One of the ways that females express bonding with each other is via vigorous genital-to-genital contact (in the scientific literature, this is referred to as "GG rubbing"). Two females approach each other and rhythmically rub together what we presume are their clitorises, for up to a minute. Their clitorises become engorged, and sometimes the participants shriek. Frequencies vary, but some observations indicate that they do this about every two hours. In bonobo culture, this female–female sexual interaction is far from unusual. It's also one of the main ways that females ingratiate themselves with new social groups.

Bonobos are surely the horniest species in all creation. GG rubbing isn't limited to females. It occurs in every possible combination, regardless of sex, age or even sexual maturity. Females also do it with males, males do it with other males, both do it with infants. Males tend to do GG in a mounting position, that is, not face-to-face, with their tumescent penises touching. Sometimes they will fence face-to-face with their erections, typically while hanging from branches on a tree.

Statistics for human sexual behavior involves a fair degree of guesswork, but I think it's a reasonable assumption that anyone who has sexual contact with multiple people many times a day is unusual. Yet for the average bonobo, that's par for the course. Yet female bonobos get pregnant and have offspring at about the same rate as chimpanzees—one child every five or six years. A rough calculation: assuming ten sexual encounters a day for five years (which is well within observed behavior), and one child in the same time period, means that about one in 18,250 sexual acts results in a baby. This is not quite the same statistic as the one quoted earlier, that only 1 in 1,000 sexual encounters in human apes that could result in a baby actually does—we're working with incomplete data-sets here. It does indicate though that we share a pattern of behavior with our nearest cousins that our fictional alien scientist might eventually spot: we have clearly separated sex and reproduction.

Much has been made of the sex lives of bonobos, understandably, as they are close evolutionary relatives, and they do have sex in ways that are perhaps more comparable to our own sex lives than, say, fruit bats or mole rats. Various claims have been made about their living in a "make love, not war" hippy commune based upon the sheer rate at which they orgasm. This pleasant sentiment has been made in contrast with chimpanzee culture, which is patriarchal, violent and murderous. As ever, the truth is somewhat more complicated.

Male chimpanzees physically fight for status and kill to reinforce it. This has never been observed in bonobos, where females dominate, and male status is in relation to the status of their mothers, whom they stay close to and depend upon all their lives. It's not quite right, though, to suggest that bonobos are a peace-loving ape for whom sex is the gentle answer to everything. Lethal aggression has been observed in wild bonobos, and lots of the ethology of bonobos has been done in the unnatural habitats of zoos, an artifice which may skew the results. These environments sometimes seem to create artificially super-dominant females, and they can be ultra-violent in conflicts. Some male bonobos in zoos are often without a full complement of fingers or toes, and one in Stuttgart Zoo had his penis bitten in half by two superior females.

It is inviting to apply human interpretations to animal behavior, and it is similarly tempting to suggest that the presence of these non-reproductive sexual acts in us relates to our evolutionary origins. But the evidence is not compelling. It is problematic to draw any strong conclusions that these are evolutionary in origin, and derived from similar roots to what we observe in bonobos, monkeys, dolphins, otters or tegu lizards (as we shall soon see). Bonobos are not our ancestors, and neither are chimpanzees. Often, when studies of our closest evolutionary cousins are discussed, the implication is that behaviors seen in these species explains our own. Great apes are more closely related to each other than they are to, say, otters, but are not evolved *from* one another. The three of us—chimps, bonobos and humans—have a shared common ancestor. What is truly fascinating about the bonobos is their evolutionary history. The Congo is a vast river that snakes through central Africa. Bonobos live exclusively on the left bank. Only recently have we begun to work out how they got there. We know that the branch that became the genus *Homo*—the humans—and the one that gave rise to the genus *Pan*, chimps and bonobos, separated six or seven million years ago, somewhere in Africa. There are meager fossils from this time and place but one reasonable candidate for last common ancestor would be the creature *Sahelanthropus tchadensis*, a hominin much more chimp-like than human. This is a messy time in ape evolutionary history, and there isn't a scientific consensus on quite how, where and when our lines diverged, nor indeed how clean the break was.

After a time though, our genealogical lines had truly split, and the chimps and bonobos would form a distinct branch. Just as we have reconstructed human population histories using DNA, we can work out who mated with whom—and when—using genetics, by comparing the DNA of living chimps and bonobos. It reveals that there has been no gene flow between chimps and bonobos for at least 1.5 million years—"gene flow" being a scientific euphemism for successful reproductive sex. Sediment analysis from the banks of the Congo River suggest that it is something like thirty-four million years old, and it is mighty enough to act as an impermeable barrier to most terrestrial animals and any proposed gene flow. It seems that natural fluctuations in its high and low tides during the ebb and flow of an ever-changing climate meant that around two million years ago, the tide was low enough that a small founding population crossed the Congo. These pilgrims were then forever isolated on the far shore, and in the time since they were transpontine,* all of the characteristics specific to bonobos emerged.

---

* "Transpontine" is an unnecessarily fancy word for "across a river or bridge." It emerged in the nineteenth century as a pejorative term for the types of sensational, sometimes saucy melodramas that were typical of theatres on the south bank of the Thames in London at that time. Bearing that in mind, I'm happy with the bonobos being transpontine.

This is how many speciating events occur: a small troop splits off from a larger group, but is not necessarily representative of the overall variation seen in the total population. Any species can be isolated in behavior—one group starts feeding off a tree that fruits at a different time—or in space—a one-way ticket across an otherwise uncrossable river. Once separated, they breed, and the gene pool from which this new population is founded is free to go off in its own direction. It's not difficult to conceive of slight differences in the first ancestors of the bonobos that gave rise to their sexual liberation. In chimps, estrous displays, including brightly and hi-vis swollen genitals, very clearly match when they are most fertile. In bonobos, females appear to be at peak fertility for much longer than they actually are. For humans, there are no convincing visible signs of periods of high fertility, which typically reaches its peak a few days after menstruation finishes.* The fact that bonobos have extended the cues of fertility beyond the obvious signals is a clue to our sex lives. It is conceivable that natural genetic variation that influences oestrus could have been amplified by natural selection in a founding population of the bonobos' ancestors.

Though I am cautious about overinterpretation of these sorts of similarities, this is crucial in thinking about our own evolution. We have characteristics in common with both species in the genus *Pan*, with whom we share a common ancestor long before either *Pan* or *Homo* evolved. They have diverged from each other, both

---

* Though many people have claimed there are physical and behavioral signs: these have included breast symmetry, face flushing, smell, gait, clothing choices and more. In almost all cases these studies have low sample sizes, or are methodologically flawed or questionable. One of the most famous, which generated a million unquestioning headlines, was that female lap-dancers in strip clubs received more tips when they were ovulating than at any other time during their menstrual cycle. Science relies on numbers to graduate anecdote to data, but this study featured merely eighteen self-reporting dancers over a period of about two menstrual cycles, which any scientist worth their salt would decry as woefully underpowered.

genetically and behaviorally. The genetics of the bonobos indicates that perhaps only a few small changes in a founding population seeded a radical change in behavior, and a totally different population structure—*Pan paniscus* are less violent than *Pan troglodytes* and use sexual encounters rather than violence to settle disputes and establish social hierarchy.

We do neither. Bonobos are fascinating, but they are also effectively an island species, and island species are often evolutionary oddities. For reasons of geographical isolation, they can be both genetically and behaviorally weird. That doesn't mean that the lives they live are irrelevant to understanding our own, but let's face it, bonobo sex lives are very different from ours, or even how we might wish them to be—it sounds exhausting. Sexual contact in bonobos serves a very different function in them, compared to us. Even though the rates of non-reproductive sex might be comparable, and may share a similar genetic basis in inception, the motivation is different, and the evolutionary histories are different. We don't touch each other's genitals to resolve conflicts, nor as a collegial greeting, nor in anticipation of a decent meal, at least not in polite society. Demystifying our own sexual preferences, peccadilloes and predilections is worthy of our scrutiny, but again, it may just be that it feels good.

Homosexuality

Of all the sexual acts that are possible, only one will produce babies. There isn't really a sliding scale here—conception is either possible or not. Due to the nature of organisms that require two differentiated sexes in two individuals, one guaranteed way to do sex and not have a baby is to do it with members of the same sex. In the future, there will be ways that eggs or sperm might be engineered to be genetically suitable to form a new conception. Both are cells that have reached a fully differentiated state—they are mature, have shuffled and halved their total DNA in preparation for meeting an equivalent cell to complete the full deck at the beginning of a new life. But we may soon be able to rewind some of that maturation process and then redirect either cell type to become something else, for example, a sperm rewound and then steered toward becoming an egg, or vice versa. That way, two women or two men could in theory conceive a child with half of each same-sex parent's genome.

That's not possible, yet. Until then, two men or two women together do not have the required genetic compatibility to fertilize an egg and produce a pregnancy. Homosexuality therefore is a sexual identity independent of the evolutionary imperative to reproduce.

I could quote dozens of different statistics here to address the question of how many people are homosexual—there is no consistent figure. Nor is there a consistent pattern of behavior that allows easy or clear definitions or demographics. Some people appear to be exclusively homosexual from a young age, and some exclusively heterosexual. Many are somewhere in between, in that they might be primarily one way or the other, but have had homosexual, bisexual or heterosexual experiences or thoughts once, sometimes or regularly. Some studies have shown that 20 percent of adults have been sexually attracted to members of the same sex, though the percentage of people who have enacted same-sex encounters is typically half of that.

Precision in these demographics doesn't really matter when thinking about the broad sweep of evolution. Homosexuality exists, and hundreds of millions of people identify as homosexual. Conception remains an impossibility following homosexual sex, which superficially suggests that it might be maladaptive. That poses a potential problem when searching for an evolutionary exploration of a particular behavior. How can sexual behavior that cannot produce offspring persist at such a high frequency? Could this be an example of something that has delineated a boundary between human animals and non-human animals?

Apparently not. Homosexuality abounds in nature too. Some of the examples have been mentioned already, though perhaps bonobos are not a great comparison, as they have sexual encounters with all members of a group all the time for complex social reasons, rather like the English chat about the weather.

Consider the giraffe. Giraffes are beloved of evolutionary biologists for a number of reasons. They are, of course, the tallest of all living animals, and that elegant neck is the primary reason why. Its exaggerated form was historically given as an example of how evolution might occur under the auspices of a now abandoned theory. Jean-Baptiste Pierre Antoine de Monet, Chevalier de Lamarck, was not the first person to consider the concept of evolution—simply, how animals change over time—but he was one of the first to think, write and publish seriously on the matter. Giraffes, or *camelopards*[*] as they were referred to in the nineteenth century, were a big part of his scheme. In 1809, the year that Charles Darwin was born, Lamarck published *Philosophie Zoologique*, in which he espoused his theories on why animals change over time. The giraffe, he argued, became "gifted with a long flexible neck" by stretching to reach the juiciest acacia leaves.† In doing so, something akin to a "nervous fluid" would flow into the neck, which would grow in response. This incremental acquisition of length would be passed on to its children and the process repeated.

Fifty years later, *On the Origin of Species* was published, which fully supplanted the idea of the inheritance of acquired traits:‡ the experiences of a life does not change DNA in a way that can be passed on to the next generation, and so has little or no influence on the genes on which natural selection acts. Darwin relegated Lamarck to the category of important, meticulous, great scientific thinkers who were not right on their big idea. We sometimes scoff at Lamarck for being wrong these days, which is an appalling slight on his thoughtful work. His ideas were supplanted by the greatest of all scientific theories, by the greatest of all biologists. All scientists need to be wrong as often as they can be, for that is the place

---

* The ancient Greek for giraffe is *kam lopárdalis*, from the "say what you see" school of biological nomenclature: camel because it has a long neck, and leopard because it has spots.

† Charles Lyell, writing in criticism of Lamarckian evolution, in Volume 2 of *Principles of Geology* (John Murray, 1837).

‡ Epigenetics is an important part of genetics. It's one of a few mechanisms by which DNA receives instructions from the environment, but in recent years, it has become fashionable to question whether epigenetic stamps can be passed on to offspring after having been acquired during life, and thus might be a kind of neo-Lamarckian evolution. There is some evidence that some epigenetic traits are passed on, but none that these are permanent. Thus, there is no evidence that Lamarckian inheritance is correct, nor has any impact either on natural selection or the robustness of Darwinian evolution.

from which we discover what is correct, and inch ever closer toward truth. There is a statue in the Jardin des Plantes in Paris, where his daughter is seen addressing an aged, blind Lamarck. The engraving on the plinth says: "Posterity will admire you, and she will avenge you, my father."

It was data that killed Lamarck's theory of evolution. There are many reasons why the inheritance of acquired traits is wrong, primarily because we have never discovered a mechanism by which the information could be passed on to subsequent generations. We don't actually see the modification of a trait in subsequent generations as a result of experience—polar bears in zoos remain white despite not spending a lot of time in snow. More prosaically, giraffes mostly forage at shoulder height, and stretching to eat higher and theoretically juicier leaves is not borne out by observation. Nevertheless, its neck is still wonderfully instructive for Darwinian evolution. Betraying its shared ancestry with other animals, it has the same number of vertebrae as both us and mice. Each one is, of course, very much bigger. That neck is also home to the recurrent laryngeal nerve—found in us and in much more distantly related fish—which innervates parts of the larynx. In giraffes, this nerve takes a preposterous fifteen-foot detour, a meandering loop around a major artery flowing directly from the top of the heart. Which is exactly what it does in us, only the length of the giraffe's neck has stretched this loop all the way up and down rather wastefully. The fact that its anatomical position is exactly the same in us and them is a stamp, a hallmark of blind, inefficient evolution in nature, which Darwin himself described as "clumsy, wasteful, [and] blundering."

The origin of that beautiful neck has also been attributed to sexual selection. It is extravagant and slightly absurd, like a peacock's tail, so might be one of those runaway traits that we see exaggerated in males of so many sexual beasts. This is where the sex lives of giraffes gets interesting. The neck is certainly a major part of sexual and social behavior. Since 1958, the male-to-male wrestling that giraffes are often seen engaging in has been called "necking." They curl their necks around each other and rut. It's incredible to watch, the necks twisting and bending at almost right angles, the normal grace of these animals replaced by ungainly aggression and awkward legs, with none of the elegant power of two stags clashing antlers.

The giraffe's recurrent laryngeal nerve

Necking, as with its human teenage counterpart, is often fore-play to some more serious sex. It looks similar to many male-to-male competitive behaviors that precede copulating with a female. They battle, and one comes out on top. The primary difference in giraffes seems to be that after a bout of heavy necking, the males will often have penetrative sex. As with so many of the interesting behaviors of wild animals that we observe and try to understand, there hasn't been a great deal of work in this area. Numbers therefore are not huge, and robust conclusions are elusive. But it does appear that the majority of sexual encounters in giraffes involve two males necking, followed by anal sex.* Not all necking encounters result in attempted or successful mounting, but in many cases, the necking males spar with erect unsheathed penises.

Giraffes tend to segregate by sex most of the time. The necking behavior happens almost exclusively in male herds. In one report, recording more than 3,200 hours of observation over three years in national parks in Tanzania, sixteen male-on-male mountings were seen, nine of which featured an unsheathed penis. The naturalists assumed initially that this was an expression of dominance, but saw no activity (normally indicated by submission, or a particular posture) surrounding the act that supported that idea. In the same period, they only saw one male mount a female. Sixteen out of seventeen is about 94 percent.

We don't know why they behave like this. Twenty-two calves were born in the same period, presumably following heterosexual action, so it follows that most mountings must have gone unobserved, but that also implies that more same-sex male mountings also happened. This data and other observations suggest that male

---

*   It's even mentioned in the 2000 film *Gladiator*: Roman animal and human slave trader Antonius Proximo, played by the late Oliver Reed, bitterly complains to another salesman that his livestock are not reproducing, hissing the line "you sold me queer giraffes."

giraffes don't have sex with females very often. When they do, they lick and smell the urine of the female, and then follow her around for a couple of days. The females will repeatedly frustrate the attempted mount by a male via the impressively nonchalant tactic of simply walking forwards. They eventually stand still if they are in the mood.

Even with scientific caution in play, it seems safe to say that most giraffe sexual encounters are male homosexual. Logic dictates that a species that is exclusively homosexual will not survive for very long. However, one in ten is still enough for a species to continue, and indeed twenty-two calves born in a three-year period is a decent brood. Female giraffes appear to be fertile and receptive for only a couple of days a year, and with a gestation period of up to one and a quarter years, they're not particularly prone to a quick generational turnaround. The homosexual encounters are clearly an activity that has some social meaning, though it's not obviously the establishment of a hierarchy or dominance. We don't know much more than that.

Many other animals also engage in homosexual sex, including rats, elephants, lions, macaques, and at least twenty species of bat. There are fewer documented examples of female homosexuality, but then there is much less data on female sexuality in humans and other animals in general. As with so many areas of science, there has been a historical skew toward understanding male behavior. Of the sapphic relations we do know about, we have a better understanding of biological principles that might be at work. Farmers are entirely untroubled by homosexual activity in goats, sheep, chickens, and even use cows mounting each other as a good sign that they are fertile. Whiptail lizards can reproduce using parthenogenesis, the virgin birth that we also see in Komodo dragons, and female-on-female mounting may be a mechanism to induce ovulation. Like bonobos, hyenas live in a matriarchy. Females are dominant, more aggressive and more muscular than males. They also have an unusual set of genitalia: the clitoris is huge, erectile and only slightly smaller than the male penis. Females engage in clitoral licking frequently, to bond socially and to establish hierarchy.

Homosexuality does pose an evolutionary puzzle, though there are plenty of ideas as to how this behavior might persist through time. In humans, there has been some evidence of regions of DNA that associate with male homosexuality. This is not a "gay gene" as the media would have you think, as there are no genes "for" complex behaviors. Rather, it seems (though data is somewhat limited) that certain sections of genetic code occur in versions that are more frequently associated with homosexuality than by chance. If that sounds mealy-mouthed and painfully caveated, that is where we are currently at with genetics and complex social behaviors. Almost no human traits are determined by the flick of a DNA switch, but instead by many genetic factors interacting and contributing small effects in concert with the lived life of experience.*

Outside of straight genetics, there have been plenty of twin studies looking at homosexuality in men. Identical twins have (very nearly) identical DNA, so any behavioral differences are likely to be caused by non-genetic, that is, environmental, factors. The various studies have produced a raft of different percentages, but all suggest that if one identical twin is homosexual, the other is on average more likely to also be homosexual, compared to fraternal twins. There's also the fact that studies show that having a homosexual older brother increases the chances of a younger brother being homosexual.

---

* As with so much research on humans, more is known about men than women, and specifically there is a dearth of research into the genetics of homosexual women. Lesbians, on average, tend to be more fluid than gay men, though, meaning that their sexual behavior and identity is more likely to change during their lives.

There is little doubt that homosexuality has a genetic component—all behaviors do. Many genes influence biological traits in concert with the environment in which they act. Genes that reduce reproductive success are eventually deleted, as the individuals who carry them will be outcompeted. The question for evolutionary biologists is why haven't those genes driven themselves out of the gene pool? Homosexual males are less likely to have children, therefore at first glance the genes involved should be prone to purging from the genome.

The first potential answer is that exclusive homosexuality may have been historically rare. There is a terminology issue here, because we tend to look at sexual behavior through a modern and Western lens. The way we typically talk about homosexuality today tends to be representative of an identity, rather than simply a description of behavior. I've fudged that boundary in these pages, but cannot when talking about homosexuality in humans. Here, I'm talking about what we might today call gender or sexual non-conformity. Having sexual relationships with people of the same sex has not always been regarded in the same terms that it is today in our culture, and in many examples it may be better thought of as "something they did" rather than "something they are." With this in mind, same-sex activity is described as occurring in the ancient Greeks, Romans, indigenous people of the Americas, Japan and many other historical societies, with a mixture of cultural acceptance.

In many of these examples, it may not have been an exclusive practice, and therefore procreation and perpetuation of a genetic basis for sexually diverse behaviors can persist without hindrance. Though homosexuality in animals is everywhere, it is also rarely exclusive. There are some cases of animals only being interested in same-sex partners: around 8 percent of domesticated rams appear to only have sexual relations with other rams. Multiple ideas have been suggested to explain this, and as so often is the case in science, the answer may well be a combination of all of them.

One of the key ideas in evolutionary biology is kin selection. It's predicated on the notion that the gene—rather than the individual, group or even species—is what is being selected by nature. It just so happens that the best way for a gene to survive into the future is by conspiring in concert with a whole load of other genes with the same selfish motivation, all kept inside a body whose job it is to secure the reproduction of those genes. This is a cast-iron theory, a towering cornerstone of evolution, and it explains the behaviors of all sorts of social organisms, especially the bees, ants and wasps, the males of which mostly don't get to reproduce at all. They do share all of their DNA with their mother, who reproduces by the bucketload, and so what has evolved is a system by which, mathematically, sterile males assisting a fecund female accommodates the survival and genetic propagation of both.

Kin selection has been suggested as a mechanism by which homosexuality might have continued in evolutionary history, despite its appearance of being maladaptive. There are two types of kin that have been explored in attempts to explain the persistent existence of homosexual men. The "gay uncle hypothesis" suggests that having a close family member who is a homosexual male will increase the survival of a nephew or niece by helping to raise, protect and nurture. The biological imperative is that they share a high proportion of their genes, and the gay uncle's genes will survive regardless of not having children himself. It's not dissimilar to other examples of reproductive support in sexual organisms, where individuals with shared genes aid the survival of offspring which are not their own. Gay uncles are similar in that regard to another possible example of family member being important for evolution: the "grandmother hypothesis" (which also serves as a means of explaining the existence of the menopause). As women become post-reproductive, they don't just shuffle off and die, but instead stick around and may assist in the raising of their grandchildren, with whom they share one-quarter of their DNA. It's a popular idea, and may be true, though data in humans is not very rich. It may also be true in killer whales, who operate complex social structures headed by older matriarchs, and are one of only three species that are known to have the menopause (short-finned pilot whales are the third).* The gay uncle idea is the homosexual equivalent to the grandmother hypothesis. The trouble is that there simply aren't many data points that support either.

There's another explanation for which the data is more persuasive. In 2012, one study indicated that the grandmothers and aunts of homosexual men had significantly more children than the grandmothers and aunts of heterosexual men. The increase in fecundity of these women appears to adequately compensate for the absence of fecundity of the men themselves. It suggests that a genetic basis that predisposes men toward homosexuality may also be the same code that facilitates increased fertility in their female relatives. It doesn't necessarily mean it's causing either, but may be tipping the scales in those directions, which is mathematically enough to compensate for the apparent loss of genetic legacy. It's an interesting idea and the data is compelling, but it's early days in this research. Though the sample size is ample, it is only one study, and much more work is needed. Whether this is the case in the exclusively homosexual rams is yet to be investigated.

---

* Much of this is derived from a decades-long study of one particular pod of orcas in the Pacific Northwest, headed by an individual known as Granny. In forty years, she didn't bear a single calf, but long-standing census data on this population indicates that a male's survival is severely impinged by the death of his mother, and radically so if she is post-menopausal. Granny died in 2017, after a long, powerful life.

In animals, homosexuality is rampant. The important thing to note here is that we don't know why giraffes or any animal engage in homosexual behavior, but we shouldn't assume that it is for reasons relevant to human sexuality. Even within human behavior, there are many examples of sexual acts between males that are ritualistic, rather than conforming to the sexual identities of men who describe themselves as gay. The Sambia are a tribe in the Eastern Highlands of Papua New Guinea who believe that semen ingestion is an essential rite for the passage into manhood. Boys approaching pubescence practice oral sex on an older man for many years, until the boy pairs with a young woman, who also performs fellatio on him for a period of a few years. Some men abandon their same-sex practices then, and others do not. Anthropologists have asserted that the male homosexual behavior is purely ritualistic and therefore not erotic, though this seems a weak argument to me, given that sexual arousal is a prerequisite for the act of releasing semen.

Men of the Marind-Anim people of New Guinea enjoy anal copulation with other men throughout their lives, associated with their belief that semen has magical properties: it is spread on arrow tips and spears to help them find their target, and ingested in concoctions by men, women and children. Taking in semen from anal sex is viewed as a means of increasing masculinity.

All the myriad versions of genital engagement in us and other animals shows that sex is very clearly not only for making babies. We sometimes make the error of assuming that, a behavior is an evolutionary antecedent to our own, or that, conversely, it has emerged in parallel because it is a good trick. The marvelous carnival in nature shows that sex is important, and that evolution finds ways to utilize what it has available to do what needs to be done. Many people know of the maxim said by the biologist François Jacob describing natural selection as a tinkerer. I like to think of the words of the US president Teddy Roosevelt: "Do what you can, with what you have, where you are."

Evolution invented parts thrown together through error and trial, which can then be deployed to try new things to fit the ever-changing environment. Sexual reproduction is clearly a useful ability to have in one's armory,* and has been with us for at least a billion years, from a time when complex life had yet to fill the oceans, the skies and the land. Since then, the basic function of making offspring from two parents has been co-opted countless times to further create endless opportunities to enhance survival. We could try to deconstruct the ontology of homosexual behavior in us. We could attempt to deconvolute and extract the biological and social cues that lead to a person having a preference or even a "type," whether it involves blondes, or kindness, or athletic physiques, or kind blonde athletic types of the same sex, or even the Papuan cultural rites of passage into manhood. Like all behavior, sexuality is programmed, not simply by genes, or the environment, but by inscrutable interactions between biology and experience.

There is a political point that unavoidably emerges from this. Homosexuality abounds in non-human animals. Superficially, it appears to run counter to the general principles of evolution, but the more we look at the ethology of sex, the more this doesn't necessarily seem to be problematic for science.

Rather hilariously, in November 2017, a Kenyan official responded to reports and photos of two large male lions in Maasai Mara engaging in anal sex (as they frequently do), with the

---

* Though we are not entirely sure why. Sexual reproduction is twice as inefficient as asexual. We know that shuffling genes via two sex types is a good way to outrun parasites that might otherwise reduce a creature's ability to survive, but the math doesn't quite add up yet. It's been a problem in evolutionary biology for decades, and is still one of the most exciting things to research.

statement that they must have copied it from watching men do it.* Imagine what he'll think when he finds out about the giraffes. Amusing though that might be, homosexual men and women are persecuted, jailed, tortured and murdered in many countries all around the world, including Kenya, and suffer prejudice everywhere. Historically, the assertion that it is *contra naturam*—against nature—has been made in order to justify that persecution. Whatever the nature of homophobic bigotry, science is not on your side. As we have seen, homosexuality is natural, and it's everywhere.

---

* Ezekial Mutua is the Chief Executive of the Kenyan Film Classification Board, the moral arbiters of movies in Kenya. In the same interview, he clarified his position: "We do not regulate animals," he added unnecessarily.

And Death Shall Have No Dominion

Briefly, let's get into one last sexual act that has a precisely zero percent chance of resulting in conception: necrophilia. There is not much data available on its prevalence in thought or deed in humans (many more people fantasize about sexual relations with dead people than actually do it), but it is illegal in most countries. The status of necrophilia under the law is nevertheless globally uneven: it was only specifically outlawed in the UK in 2003, and in the US there is no federal legal edict on necrophilia; each state has its own particular view. Sex with the dead is considered to be a paraphilia, an unnatural deviance indicative of an abnormal psychopathology.* That's a fairly uncontroversial thing to say, and yet this practice is seen in dozens of animals.

Zoo behavior is often weird, the artifice of a captive life driving deviations from what animals in their natural habitat might normally do if left unbothered by humans. Nevertheless, many animals in captivity have engaged in activities that perhaps visitors were not counting on seeing when they went to the zoo, such as

---

* A 2009 report in the *Journal of Forensic and Legal Medicine* described a new classification system for necrophilia, with ten classes, including role players, who get sexual pleasure from pretending their live partner is dead; romantic necrophiliacs, who in bereavement remain attached to their dead lover's body; opportunistic necrophiliacs: those who normally have no interest in necrophilia, but take the opportunity when it arises; and homicidal necrophiliacs: people who commit murder in order to have sex with the victim.

male pilot whales who have been seen attempting penetrative sex with dead females since the 1960s.

It's not only something that occurs in the unnatural context of captivity though. Necrophilia is common in the wild. Sex with dead individuals has been known about in Adélie penguins since the earliest days of Antarctic exploration, as documented by the scientist aboard Captain Scott's last and fatal venture south. The penguin's behavior was deemed "astonishing depravity," and far too unsavory for delicate Edwardian sensibilities; it was redacted from the larger report released to the public, written in Greek, and made available only to a select group of stout-minded British gentlemen scientists.*

In 2013 in Brazil, two male tegu lizards of the species *Salvator merianae* were observed copulating with a dead female for two days, during which time she had bloated and begun to putrefy.† Such is the biological imperative, probably driven by the enduring presence of pheromone signals that females emit to indicate sexual availability, that male frogs and snakes have been recorded attempting copulation with females who have been decapitated and run over by a lorry, respectively. In a brutal write-up published in 2010, male sea otters were observed repeatedly and successfully attempting forced copulations on females, sometimes drowning them, and at other times causing injuries (such as perforated abdomens and vaginas) so severe that the females subsequently died. The males were then seen copulating with the carcasses for several days. Even

---

* The paper was authored by the scientist George Levick on Captain Robert Falcon Scott's 1910–12 expedition, which ended in Scott's death. Levick wrote of the young male penguins as "hooligan bands of half a dozen or more and [that] hang about the outskirts of the knolls, whose inhabitants they annoy by their constant acts of depravity."

† Full respect to the author Ivan Sazima for entitling his report "Corpse bride irresistible: A dead female tegu lizard (*Salvator merianae*) courted by males for two days at an urban park in southeastern Brazil."

more startlingly, they did this not only with otters of the same species, but also with harbor seals.

Perhaps there isn't a more pertinent moment to again point out that behaviors we see in non-human animals are not necessarily related to our own. Whatever the pathologies that drive necrophiliac behavior in humans, they are unrelated to the motivations of other animals, about which we can speculate scientifically or remain agnostic.

Necrophilia, while being distasteful, has turned out to be essential for understanding some sexual biology, via careful experimental design. Earlier, I mentioned sperm competition as a significant mechanism by which males compete for females by engineering conflict not at the level of the individual, but within their ejaculate. In some bird species, males seem somewhat unfussed about whether their mate is actually alive or not, and scientists have used this lack of discernment to study sexual biology. Researchers find recently deceased female birds, and glue them to a branch. Males mate, and deliver their sperm via the dignified cloacal kiss, and fly away, their biological imperative apparently fulfilled, only to have scientists retrieve the ejaculate for laboratory analysis.

Sex and Violence

Sex is a physical act between individuals, and for reasons discussed earlier, sexual proclivities may not be evenly matched in males and females—the metabolic investment in eggs and sperm is uneven, and this mismatch drives sexual selection. From this evolutionary force, we see obvious physical differences between males and females, such as in size, in genitalia (the primary sexual traits), in ornamentation (secondary sexual traits), and in behavior. The fact that there is a mismatch in sexual imperatives and that sex is a bodily act means that physical conflict is a frequent part of sexual encounters.

The wording in that paragraph is deliberately cautious to the point of being inelegant. The language we use in describing sexual behavior in non-human animals is problematic. We have specific words for specifically human behaviors, but ones that seem to have very clear analogues in non-human animals. There are several examples of "transactional sex" in, for example, female Adélie penguins, who need stones to build their nests, have sex with an unattached male, and take a pebble from their stash afterward. This gets referenced as "prostitution" in media coverage. There is one study in which rhesus macaques appear to trade a commodity, in this case water, in exchange for simply looking at images of higher status monkeys, and photos of the genitalia of estrous females from behind; in the media, this was reported as "monkeys like pay-per-view pornography."

Sex in animals frequently appears violent. We must tread carefully here. Sexual violence in humans is a most serious crime and rape an act of profound violence and violation of personal autonomy. But it is also as ancient as culture, with descriptions of sexual violence and rape in our oldest texts. These include the rape of Hera, Antiope, Europa and Leda, all by Zeus; Persephone by Hades; Odysseus by Calypso; in Genesis in the Hebrew Bible, Lot offers his two virgin daughters to be raped by an angry mob, though they turn down his offer, and are blinded by angels instead. The angels burn Sodom, and Lot and his family flee, but his wife did look back, and she was turned into a pillar of salt.

Rape has been suggested by some psychologists as an evolutionary strategy for humans.* To my mind, this is ill-considered speculation, perhaps the most destructive and controversial version of a "just-so story." Regardless of the significant social implications of rape having a direct evolutionary benefit, purely from a scientific view, it struggles to get beyond the point of being nothing more than a suggestion, primarily because the data to support it is so inept.† The idea, as with much of evolutionary psychology, is that today we see the vestigial remnants of a behavior that evolved and was promoted by natural selection in our prehistory: men who raped in the Pleistocene sired more children than those who reproduced with consent, and therefore genes that encourage coercive sex would propagate, and continue to this day. A couple of the arguments for the case go like this: rape victims tend to be

---

* Most prominently in the book *A Natural History of Rape: Biological Bases of Sexual Coercion* by Randy Thornhill and Craig Palmer (The MIT Press, 2000).

† For the purposes of this short discussion, I am referring only to rape of women by men. This is the crime for which we have the most data. Male-to-male rape (and to a lesser extent, female-to-female rape) does of course occur, but cannot result in conception, so there is no cogent evolutionary argument for their existence.

young and in their prime reproductive years, therefore the selection of victims by males is to maximize the chance of a pregnancy; second, women of this age are more likely to fight against the rapist, suggesting that they have more reproductive investment to protect in their own autonomous mate choice, which indicates they are more desirable targets for rapists.

These are terrible arguments, evidence-free assertions that waft away in the gentlest breeze. Problems with the first are manifold. Rape is one of the least-reported crimes: statistics vary, but they suggest that almost all rapes are undocumented in official crime statistics, because victims do not go to the police. In the UK in 2017, for example, the estimate is that only 15 percent of rapes were reported. These numbers make it virtually impossible to assess a pattern that fits the claim that rape is primarily an attack by men on women at their most fertile, which is central to the thesis. Plenty of rapists attack older women, presumably of a post-reproductive or lower fertility age, and there are many cases of the rape of children who cannot conceive. A significant proportion of rape occurs within marriages or long-term monogamous relationships, though robust stats on spousal rape are not forthcoming. Nevertheless, spousal coercive sex seriously undermines the idea that rape is a means of spreading genes more widely than by consensual sex. Even if any of those assertions were supported by facts, which they are not, then an evolutionary basis would still need the one killer argument in its armory, which is the measure of evolutionary success: men who rape should have more children than those who don't. We have no data on this, nor anything suggesting that it might be the case. The same advocates for rape as an evolutionary strategy volunteer their own counterargument: if rape does not have a direct evolutionary basis, then it is a by-product of evolution. This is an equally vacuous statement because, as we have seen, all behaviors are the by-product of evolution. That does not mean that they are adaptations that have been positively selected. Being talented at ice-dancing or scuba-diving has not and cannot have been directly selected by nature, and therefore are also by-products of our evolved brains, minds and bodies. Because that argument is so weak, it is implied that the natural selection argument is strong. But it's not. Arguing for rape as having a natural history directly rooted in a biological strategy might be the nadir of evolutionary psychology. In this case, mocking it by calling it a "just-so story" undermines its intellectual speciousness.

All of this poses many problems when we turn to the sexual behavior of other animals. There are many examples of coercive or apparently forced sex in animals, but the question of whether we can describe any of these as rape is difficult. Rape has a specific legal meaning, one that in most definitions includes an absence of specific consent from the victim. As such it is therefore specific to human animals, and we should be very cautious about applying the word "rape" to any other species, as we cannot necessarily apply the concept of consent to a non-human animal.

Coercive sex is, however, common, as is aggression by males on females (the opposite is rare). Apparently forceful mating is seen in animals from guppies to orangutans. Females show resistance to copulation, which is ignored by the males. Chimpanzees will bite, charge, screech and wave branches at females as a tactic to force them to submit.

There are other, subtler coercive tactics. Female and male newts engage in an embrace-wrestle called "amplexus," as do many external fertilizers, but for the red-spotted newt *Notophthalmus viridescens* this is not the sexual act itself. Instead, the male follows this wrestle by depositing a spermatophore, a little packet of sperm, which the female either chooses to pick up, or runs away. But males try to tip the balance in their favor by rubbing hormonal secretions on the female's skin during amplexus, which has the effect of making her more likely to ingest the sperm. The males effectively drug the females.

Another tactic is referred to as intimidation, though sexual bullying might be more appropriate. Female water striders of the species *Gerris gracilicornis* have covered genitals unlike many other striders, a kind of natural chastity belt. They can only engage in intercourse with a male by revealing their genitals willingly. This physical hurdle has evolved to prevent the coercive sex whereby males can simply wrestle and mount a female, who then has to fight them off, which is tiring, or submit. Therefore, for *gracilicornis*, coercive sex should be off the menu. But, as we often say, evolution is cleverer than you, and in this case much more devious. Male striders tap the surface of the water at a particular frequency that draws the attention of another insect, greater water boatmen, to the females. Water boatmen eat water striders. The females respond to this threat by allowing mounting of the male, who can stop tapping and thus call off the potentially lethal attack.

These acts of coercion show the staggering lengths that evolution has gone to in order to manage the genital arms race between males and females. Again, while they may look similar to human behavior, and the language we use is familiar to us, these comparisons cannot be thought of as homologous. In general, females are choosy and males indiscriminate, and the math of these strategies explains them. It has to: harassment, intimidation and actual violence all have a cost to the females, and this cost may reduce her overall reproductive fitness, whether it is via injury, increased risk of being eaten or simply the loss of time to have sex with preferable males. Whatever the strategy, the female tactic is generally to have evolved traits that reduce or minimize the cost of coercion.

Not all of this violence in sex is easy to explain. In the case of the sea otters described earlier, it is reasonable to assume that female otters do not wish to be penetrated so vigorously that they die; that is an evolutionary strategy that is difficult to explain. Here, the cost to the female is ultimate, but the male does not benefit either: a dead female will not conceive, and his genes will not endure. Puzzlement at this behavior is increased by the fact that the male otters also perform the same violent acts on another species, harbor seals, for whom there is no chance of their becoming pregnant, and they suffer similarly lethal consequences. The killing is perhaps explicable as both species may compete for resources, but the copulation with their carcasses is perplexing.

Nature is "red in tooth and claw," as Alfred Lord Tennyson wrote in the poem *In Memoriam A.H.H.* I expect he wasn't referencing newts or the water strider. In that now famous line, Tennyson was considering, pre-Darwin, the apparent callousness of nature. Nature is not cruel, it is simply indifferent, and these behaviors show a disregard for other living things, rather than malice. Humans alone are capable of cruelty, and sexual coercion and rape are immoral and criminal acts. Describing non-human behavior in these terms trivializes rape.

We do need to talk about dolphins though, as their sexual behavior is concerning and much discussed. We have a strange relationship with dolphins. We are often in awe at their intelligence and grace, and the tricks they do for us in captivity and in the wild; and they have nice smiley faces. Dolphin is a loose and informal name for several different groups of cetaceans, including the delphinidae (ocean dolphins), and three classes that live in rivers or estuaries (Indian, New World and brackish).* They are smart and have large, complex brains (see page 37), and complex societies, notably (but certainly not exclusively) the bottlenose dolphins best studied in

---

* The baiji, or Yangtze river dolphin, represents one other genus, but the last one was seen officially in 2002. One possible sighting in 2007 was unverified, though this doesn't change its status as functionally extinct. Maybe it was the last, or second to last, it doesn't really matter. When a population is reduced to one or even two members, there is no hope for the survival of the species.

Shark Bay in Australia. Two or three males will form a gang that swim and hunt with each other, called a "first-order" pair or trio. Sometimes, two pairs will team up—a second-order alliance.

These Shark Bay dolphins are also viciously violent. When breeding season comes around, there is fierce competition for access to females, as happens in many sexual species. Mostly in nature, that competition is between individual males. The bottlenose dolphins have a different tactic: they form gangs. Alliances are an essential part of the mating strategies of the males: first-order partnerships will single out a female, rush at her, and then herd her away to have sex, which is coercive (this is a general assumption, because it is rarely seen). During this aggressive corralling, the females repeatedly try to escape, and do so in about one in four attempts. The males restrict their attempts at freedom by charging in, and bashing her with their tails, head-butting, biting and body slamming her into submission. Second-order alliances do the same but the team-up makes a ratio of five or six males to one female. The males are often closely related in these alliances, so as a means of transferring their genes into the future, this fits perfectly well within evolutionary theory. On occasion, they form looser "super-alliances," where multiple second-order gangs will join forces—up to fourteen individual males—to corral a single female. These gangs don't tend to be closely related.

It should be noted that forced copulation has not been directly witnessed, as far as I am aware. The evidence comes from observations of the pre-copulatory behavior, and physical evidence of violence on the females. Many people talk semi-jokingly that in contrast to their cute and smart image, dolphins rape. There is no doubt that sexual coercion is part of their reproductive strategy, as it is in many organisms, and the behavior is violent, but we must be careful not to anthropomorphize their behavior, cute, smart or horrid.

Infanticide is another unpleasant behavior seen in dolphins, which also gets translated into murder in the popular press, though it should be noted that in plenty of other organisms, both males and females kill the young of others within their own species as a reproductive strategy. A female lion lactates for more than a year when she is nursing cubs, and during this time won't breed. Males acting alone or sometimes in packs will kill her young in order to bring her back to being fertile so they can then sire a pride. Mother-and-daughter teams of chimps in Tanzania have been seen killing and eating the babies of other parents for reasons that are not clear. Alpha female meerkats will kill the litters of subordinate females so that they will be free to help nurture the alpha's litter. Female cheetahs get around all these issues by copulating with multiple males, whose sperm mixes internally, and each of her litter will have a different father.

There are plenty of reports of dolphin calves washed up on beaches with extreme injuries. One study in the 1990s reported nine that had died of blunt-force trauma, including multiple rib fractures, lung lacerations, and deep puncture wounds that were consistent with bites from an adult dolphin.

Are dolphins murderers or rapists? No, because we cannot apply human legal terms to other animals. Is the behavior distasteful to us? Yes, but then again, nature does not care what you think.

This walk through some of the grimmer aspects of the behavior of animals serves as a reminder that nature can be brutal. The struggle for existence means competition, and competition results in conflict and sometimes lethal violence. We recognize these behaviors because humans compete and can be horrifically violent. But we are not compelled to act violently. The evolution of our minds may have gifted us the ability to craft tools that enable massacres. But it has also furnished us with choices unavailable to our evolutionary cousins. We are different because, with behavioral modernity, we have eased our own struggle for existence away from the brutality of nature, so that we are not obliged to kill others or force sex upon females in order to ensure our survival. The question is, how did this happen?

PART TWO

The Paragon of Animals

Everyone Is Special

In *The Descent of Man*, Darwin considers the difference between the minds of humans and other creatures. He speculates on the cognitive abilities of a hypothetical ape, and notes that while it could crack a nut with a stone, it would not be able to fashion a tool from that stone. Nor could that ape "follow out a train of metaphysical reasoning, or solve a mathematical problem, or reflect on God, or admire a grand natural scene."

But Darwin goes on to suggest that "emotions and faculties, such as love, memory, attention, curiosity, imitation, reason" may be found in some incipient form in other animals. He writes that the difference between the mind of humans and other animals is one of "degree and not kind."

This section is a beautiful piece of prose, and that is a memorable phrase, which has been appropriated by fields well outside of biological evolution to describe things which differ not fundamentally, but merely by position on a spectrum.

As for its original meaning in the evolution of us, I no longer know if it is true. As we have seen, in technology, in sex, in fashion, we are different from other animals. But the implication that the differences between us and them are determined by our relative position on a line is questionable. Our use of tools is so very much more sophisticated than that of a crow or dolphin, or even a chimpanzee, that it doesn't seem fair to simply attribute this to our more advanced position on a spectrum. Our sexual desires and proclivities may resemble behaviors in some animals, but the motivation for the rampant sexual behavior of the bonobos serves a very different social function, even if the various physical mechanics are familiar to us. On the other hand, maybe our enjoyment of oral sex is not really different from those two unusual bears in Zagreb. Is Julie's ear-twig merely a simpler, incipient version of any extravagant fashion that we adorn ourselves with today?

We have a culture that doesn't only surpass all others in sophistication, it simply doesn't exist in any other species, and the way in which our knowledge is passed around, among contemporaries and down the generations, is barely glimpsed outside of the genus *Homo*.

Perhaps the phrase "different in degree and not kind" is too simple, too binary to be of use in understanding the story of the human animal. Perhaps it is better to revel in the complexities of our evolution, and just acknowledge, without superiority or judgment, that we are different.

How did this arise? Why are we different? We strive so desperately to find it: the switch that turned us from one being into another, the thing that made us human. In our stories, and in science to a certain extent, we crave triggers. We search for clarity and hope for narrative satisfaction, and that in our searching we will reveal the story of how we became us.

Here's the rub: evolution doesn't work like that. Even as a metaphor for some real transition in our origins, drama imposes a flip which never happened. Certainly, there are nodes in the story of life on Earth. These pivotal events are few and far between, but there are singular moments when the trajectory of evolution is forever altered. For example, the birth of complex life some two billion years ago—when one cell climbed inside another, and in so doing spawned everything else on the tree of life. That appears to have happened just once. There was only one meteorite that triggered the end of the dinosaurs' 150-million-year reign, and in doing so, freed up environmental niches in which small mammals and small birds could thrive. These are undeniable events—things were different the day after—but mostly life evolves messily and slowly, much like our own lives. It may be punctuated by moments, but you are mostly a connivance of four billion years of biology, and a few years of living in and among other organisms and your environment.

It is not easy to consider our own story. Evidence from prehistory is scarce, and we put together a narrative using fragments from our past. Two factors significantly hamper our ability to get our heads around our own evolution and existence. Time is a concept that plays against us. The timescales we think about in evolution are inconceivable and bear almost no relationship to a lived life. We can think sensibly about two, maybe three generations in either direction from us—our great-grandparents, or great-grandchildren. But when thinking about the foundations of our species, we are talking about thousands or more. *Homo habilis*, for instance, emerged more than two million years ago, a period filled with hundreds of thousands of generations.

And there's another rub. Singular moments or single causes are easier for us to process than an inscrutable system millions of years in the making. Science is inadvertently responsible for fomenting both granularity and linearity in understanding complexity, because that is how we have to pick apart intensely integrated systems such as our bodies, our minds, or our evolutionary story. We find one tooth or a hyoid bone in the dirt from an age ago and try to extract every ounce of data out of it, and then fit it back into the bigger picture of the lives of ancient people. Or we take one gene, and look at how it changed in people, and where they took that gene all over the world. Each of these elements is but one part in a giant four-dimensional puzzle; four because living organisms pass through time as well as physical space. As a species, all the things we do are unique, and are also seen all over the natural world.

And so we chip away at ourselves, perpetually introducing new ideas and new data and striving to ignore or file away the preconceptions or baggage that might hamstring our understanding of our own story.

But why *are* we different? Separating biological from cultural evolution creates a false divide between them, when in fact they are intrinsically interdependent—biology drives culture and vice versa—but it is necessary to look at the individual pieces of the puzzle before we can fit them back together. Let's deal first with the biological, which in evolutionary terms means DNA.

Genes, Bones, and Minds

Genes are the units of inheritance, the things that are selected by nature to be carried into the future. Nature sees the physical manifestation of a gene—the phenotype—and as a result of that trait enhancing survival, the DNA underwriting it succeeds, and is passed on down the generations. Genes are the templates on which our lives are built.

Our knowledge of how DNA translates into a life has been radically transformed in the last few years for two reasons. The first is our continued pursuit of how the code actually works, how genes spell out natural variation between humans, how variation is and has been spread around the world, and how faulty genes can result in disease. The code hides biological data, which itself is written in genes, which are hidden in three billion letters of DNA, spread over twenty-three chromosomes, held in the little nut in the center of most cells, the nucleus. We all have effectively the same set of genes, but each individual iteration of a gene is slightly different, and in those differences lie natural variation between people. Though we still have a long way to go, we have a strengthening grasp on how genomes work and how that basic sequence of letters becomes living biology. The closer two individuals are related, the more similar their genes will be. This applies in families, within species, and across species. Because we have the same genes, we can compare the precise differences between them, and work out if the differences are meaningful or not. US English and British English have many different spellings for the same words, but the *color grey* is the same as the *color gray* on both sides of the Atlantic, and the meaning is retained despite the different spelling. On the other hand, *appeal* and *appal* (a variant spelling of appall) are only one letter different but the deletion renders the meaning almost opposite. DNA subtly changes over time via the genetic equivalents of typos—spelling mistakes which slip through due to inaccurate copyediting by the proteins that check the code after it has been replicated. These errors accumulate at a fairly constant rate, which means that the differences between genes in individuals and species acts as a clock, which started ticking when those typos originated in an ancestor's sperm or egg, and were passed on to their children. Profound advances in genome sequencing in the last few years have made deciphering the biological code easy, cheap and quick, and now we have petabytes of DNA from millions of living humans and other animals.

The second reason why genetics has drastically changed in recent years is all of the same reasons described above, but applied to the genomes of people who have been dead for years, decades, centuries or even hundreds of millennia. DNA is a remarkably stable data storage format. In a living cell, it's preserved by active maintenance, by proteins that spell-check and edit and make sure that each time it is copied, mistakes are limited. In a dead cell, there is none of that proofreading, but DNA can persist for thousands of years, in the right conditions—preferably dry and cold, and in the presence of as few other organisms as possible. With the genes of the dead, we can reconstruct genetic relationships that are otherwise lost in time.

With those two advances in genetics, we have arrived in a new era of understanding inheritance, enabled by crunching reams and reams of data in the form of genome sequences, which can be processed to see wispy, subtle patterns only revealed with powerful statistics. With these new tools to hand, we can make progress in understanding how early humans changed into the beings that we are today.

24 − 2 = 23

Species are defined by their morphology and not by their DNA. That taxonomy exists for historical reasons: we were classifying organisms using the current system since Linnaeus devised his binomial nomenclature in the eighteenth century—genus followed by species, *Homo* and *sapiens*, *Pan* and *troglodytes*. Every human being has a unique genome, but they are similar enough that we can be sure that we are one species. Crucially, all living humans typically have the same number of chromosomes.* Each chromosome is a long thread made of DNA, and parts of each thread are genes, around 20,000 of them for us, spread over those twenty-three pairs of chromosomes. Gorillas, chimpanzees, bonobos and orangutans have twenty-four.

Chromosomes are all different sizes, and our number 2 is one of the biggest, representing about 8 percent of our DNA, and harboring around 1,200 genes. It's that big because at some point, maybe six or seven million years ago, one member of the common ancestors of all the great apes gave birth to a child with a gross chromosomal abnormality. During the formation of the egg and sperm

---

* There are a handful of viable chromosomal abnormalities where people have extra chromosomes, or in some cases too few. The most well-known is Down's syndrome—an extra chromosome 21, three instead of the requisite two—but there are also conditions such as Klinefelter's (a man with an extra X, making him XXY) and Turner's (a woman with an unpaired X).

that would fuse to begin this life, instead of replicating all the chromosomes perfectly, somehow two of them crunched together and stuck. By lining up all the great apes' chromosomes, we can see very clearly that the genes on our chromosome 2 are spread over two different chromosomes in chimps, orangutans, bonobos and gorillas.

Most mutations of this magnitude are utterly lethal, or cause terrible diseases, but this ape got lucky, and was born with a fully functional genome that was significantly different from his or her parents. From that point on, the genealogical lineage of twenty-three pairs of chromosomes would trace a line all the way to you. We now have the full genomes of other types of humans, the Neanderthals and the Denisovans, but annoyingly, chromosome count is not preserved in the fragmented DNA that we can get from their bones. We reasonably suppose that they also had twenty-three pairs due to their relatedness to us, but we cannot be absolutely sure, until we get much better quality samples out of the sparse DNA-laden bones. We know we bred with them, and a different number of chromosomes is often a very sturdy barrier to reproductive success, though not always: living equids—that is, species of horse, ass and zebra—show clear evidence of having interbred despite having chromosomes varying between sixteen and thirty-one pairs. No one has figured out how though, yet.

We haven't been able to extract DNA from most specimens from the ancient human family tree, and may never be able to, as so much of our ancestors' remains are from Africa, where heat renders preservation of DNA fairly untenable. It is likely that all apes after the split from what would become chimps, bonobos, gorillas and orangutans, have twenty-three pairs of chromosomes. Genes are translated into proteins, and proteins perform actions in bodies. This includes everything from forming hair or the fibers in muscle cells, to manufacturing the components of cells that are fatty or bony, or acting as the enzymes and catalysts that process food or energy or waste. Subtle variations in genes result in changes in the shape or efficiency of proteins, and that means that some people have blue eyes and some have brown,* or that some people can process milk after weaning, but most can't, or that some people's urine smells after they've eaten asparagus and other people's doesn't (and some people can smell it and others can't). Genetic variation becomes physical variation. We call the specific sequence of DNA the genotype, and the physical characteristic it encodes the phenotype.

DNA changes randomly, and these mutations are subject to selection if the phenotype is beneficial to the survival of the organism, or impairs it. Over time, bad mutations are generally weeded out, because they reduce the overall fitness of the creature that bears them, and good ones spread. Sometimes it's a bit of both: having one defective version of the beta-globin gene acts as protection from malaria; having two copies means you get sickle cell disease. Many simply drift—the genetic mutations encode change that is neither good nor bad.

Though we have almost the same set of genes as the other great apes, many of those genes are slightly different, and a few of them are new to the human genome. Those differences are us. There are lots of ways that, over generational time, genes and genomes can change and create new information. They can subsequently be

---

* The genetics of eye color are taught in schools as one of the benchmarks of understanding genetics. In fact, eyes are a useful benchmark of how poorly we understand inheritance. Though the brown version of one gene is dominant over the blue version, there are many other genes that have a role in determining iris pigmentation, meaning that there is a spectrum of eye colors from palest blue to almost black, and it is effectively impossible to accurately predict what color eyes a child will have based on the eye colors of the parents. Furthermore, it is possible for any color combination in the parents to produce any color in the child. Genetics is complex and probabilistic, even in the traits that we think we understand well.

selected, in a direction that may ultimately become a unique combination for a distinct species. I won't go through all of them, as all happen in all creatures. But some mechanisms by which mutation occurs are pertinent to the formation of our uniquely human genome and are worth looking at more closely.

Duplication

Imagine you were composing a symphony, and you'd written it down by hand onto sheet music, of which you have only one copy. If you wanted to experiment with the theme, you'd be crazy to write over the only copy you have, and risk messing it up with something that doesn't work. You'd photocopy it, and use that one to play around, while making sure the original was preserved intact as a backup. That's not a bad way to think about genome duplications. A working gene is constrained by being useful, and is not free to mutate at random, as most mutations are likely to be deleterious. But if you duplicate a whole section of DNA containing that gene, the copy is free to change and maybe acquire a new role, without the host losing the function of the original. That's how a primate ancestor of ours went from two-color vision to three—a gene on the X chromosome encodes a protein that sits in the retina and reacts to a specific wavelength of light, and thus enables detection of a specific color. By thirty million years ago, this had duplicated, and mutated sufficiently that blue had been added to our vision. This process has to happen during meiosis, where sperm and eggs are formed, if the new function is to be potentially permanent, as the new mutation will be inherited in every cell of the offspring, including the cells that will become the sperm or eggs.

Primates seem prone to genome duplication, and the great apes particularly. Something like 5 percent of our genome has come about from duplications of chunks of DNA, and about a third of that is unique to us. Duplicated regions of the genome have always been troublesome to analyze, simply because they are copies and look much the same as each other. But with patience and persistence, geneticists are beginning to work out how to sieve them out, and with that comes new insights into why we have so many photocopies, and if there are genes within that give us powers beyond those of our ape cousins.

So far, a handful of genes have been identified that are intriguing duplication candidates that appear to be unique to us. They've all got extraordinarily dull names. In June 2018, a subtly different version of a human gene called NOTCH2NL was unearthed from a mass of very similar ones, but crucially, this new one is not present in chimpanzees. It looks like an earlier version of NOTCH2NL was duplicated poorly in a common ancestor of all the great apes, but around 3 million years ago, the dud version was spontaneously corrected in our lineage, whereas it remains mangled in chimps. We don't know what the uniquely human version of this gene does precisely, but it appears to bolster the growth of a type of brain cell called radial glia, which span the cortex and have the job of making more neurons, and thus fuelling brain growth. As ever, we learn a lot about what genes do by studying what effect they have when broken by mutations, and one of the diseases associated with mutated NOTCH2NL is microcephaly—a reduction in brain size.

We have four copies of a gene called *SRGAP2*, where other apes have one. We can see that these duplications occurred at specific times: the first was around 3.4 million years ago; this version was then copied twice more, once 2.4 million years ago, and again a million years ago. The next thing you do is look for the tissues in which this gene is active, and this is where it gets really interesting. The first and third duplications don't appear to do much, and might be just sitting there slowly rusting in our genomes. But the second duplication resulted in a gene that does its business in our brains. It seems to have the specific effect of increasing the density and length of the branching extensions called dendrites in neurons in the cortex. This type of neural patterning is unique to humans: mice brains don't have it, but when we insert the human version into mouse neurons, they grow into fattened, dense dendrites. This version of the gene, *SRGAP2C*, emerged 2.4 million years ago, at a time when the brains of our ancestors significantly increased in size. It was also around this time that we began to flake and knap stones into the Oldowan tool set.

The connections seem obvious, but I am speculating. Though perhaps not wildly. These three things—the timing of the birth of this new gene, what it appears to do in the brain, and what behavior was emerging at that time—are temptingly related. For now, that is the best we can say. This is not the one gene that made our brains the way they are, but it might be one of a few, even if we don't know quite what they do. They become clues to isolating key differences between ours and the brains of others, and more genetic hints will emerge in time. None of them are singular triggers though, just part of the picture of how evolution crafted us.

Brand New Genes

Duplication and transfer from other genetic sources are examples of nature's ability to co-opt existing tools: evolution the tinkerer. Evolution also creates from scratch. We call these *de novo* mutations, and they arise when a seemingly nonsensical run of DNA mutates and changes into a readable sentence.

The way the code works is that there are four letters in DNA, and in a gene they are laid out in three-letter chunks—each of which codes for an amino acid—which are strung together in a particular order to make a protein. Using language as an analogy, we have letters (of which there are twenty-six), words (which can be any length), and sentences (which also can be any length). In genetics, there are only four letters, and all the words are three letters long. The gene is the sentence, and like language, these can be any length. When a gene is created from scratch, it still has to have evolved. Unlike duplications and insertions which have evolved somewhere else, *de novo* genes aren't installed into our genomes already in working order. In a book, every word should have a purpose; genomes are full of DNA that isn't words or sentences, just random bits of filler. So, imagine there was a section of letters like this:

THEIGDOGATETHEFOXANDWASILL

If you strain, you can probably see that there is a simple sentence in there struggling to get out. If we insert a B after the third letter, it becomes:

THEBIGDOGATETHEFOXANDWASILL

Which if you add spaces, three letters per word, becomes:

THE BIG DOG ATE THE FOX AND WAS ILL

It only makes sense with all the letters in the right order. In genetics, this is called an "open reading frame." There are no spaces in genes, but cells still understand the three-letter structure. *De novo* genes arise when a clump of letters is converted into a meaningful sentence by chance, and thus suddenly becomes understandable by the mechanics of the cell, and translated into a protein. The protein that results is utilized in some way. If it is used, then the organism that has acquired this new gene will pass it on.

In 2011, sixty genes that are new to humans were identified, and this number may still go up. We mostly don't yet know what they do, but they all tend to be short, which makes sense, given the way they arise—the longer a sequence is, the more chance that the open reading frame will collapse. The fact that these are unique to humans does not make them defining genetic characteristics of humans. They might not do much at all; genes that have mutated to be unique to us but are inherited or duplicated from ancestors are overwhelmingly more common in our genomes.

Invasion

One other thing to note is that genetically, we are not entirely human—around 8 percent of our genome has not been inherited from an ancestor at all. Instead, it's been forcibly implanted into our DNA by other entities trying to enact their own replication. Think of a virus as a kind of hijacker, who breaks into a factory and replaces the normal plans with their own, so that the factory starts producing according to the hijacker's wishes rather than the factory owner's. When a virus storms the barricades of our cellular factories, it brings with it its own DNA (or RNA)* and can insert it into the host genome, whereon the host cell simply does the virus's bidding and makes new viruses. More often than not, this insertion is bad news. Much of the symptoms of having a cold, or many other viruses, is our immune system reacting to an alien invasion or the cell's self-destruction at the behest of the virus. Sometimes the insertion can be in the middle of a gene that puts the brakes on how often a cell divides and, in doing so, can cause unregulated

---

* RNA is the cousin to DNA. It is a very similar nucleic acid (the –NA bit), but unlike DNA, which is normally two strands linked into its iconic double helix structure, RNA remains in a single strand. In the process of a gene becoming a protein, the process typically is that DNA is transcribed into a RNA molecule, which is then translated into a series of amino acids that form the protein itself. Some viruses store their genetic material as DNA, but others, such as HIV, carry only RNA, which is converted to DNA once it has infected a host cell, and this gets inserted into the host genome using a viral protein called an integrase.

division—a tumor. Sometimes though, they just sit there doing not much. The DNA of the virus is inserted, and it's not a big deal. It has happened countless times in our evolution, which makes up that 8 percent. In total, for comparison, that is much more DNA than comprises our actual genes, and more than several chromosomes, including the Y. By this measure, humans are significantly more virus than they are male.

What this alien DNA is doing in us varies, but one example shines above all others, and it is in the formation of the placenta. There are cells throughout the body in specialized tissues with the beautiful name syncytium. They have multiple nuclei, formed when cells fuse with each other, which happens in the development of some muscle tissue, bone and heart cells. Syncytium in the placenta make up a highly specialized and essential tissue with the even more beautiful name syncytiotrophoblast. These are the spindly fingers from the growing placenta that invade the wall of the uterus and provide the interface between the mother and embryo, where liquids, waste and nutrients are exchanged. It's also a tissue that suppresses the immune system of the mother, to stop her body automatically rejecting the growing child as an alien presence. These cells are at the junction of human reproduction, where one life is giving rise to the next. The genes that drive those placental cells to form are not human at all. Primates acquired them from a virus around forty-five million years ago; in the virus, the genes also encourage fusion of the host cell with the virus itself, and help suppress the immune response to this infection. But they were co-opted and integrated into our own genomes, and are now essential for successful pregnancy. Of course, mammals have had placentas for much longer than forty-five million years, and this is a truly weird and wonderful story in evolution. In mice, who also have an essential syncytiotrophoblast, they have a very similar set of genes involved that have also been acquired from a virus, but a completely different one to us. It is an astonishing example of convergent evolution at a molecular level. Acquisition of a viral genetic program has driven the development of mammals several times over, in almost identical ways.

Hands and Feet

We have duplications of genes in combinations that are unique to humans. And we have versions of genes that are also found only in us. We can also talk about what specific genes are doing in us.

We've compared the behaviors of animals with us in this book, and we can extend this to genetic comparisons too. We have many genes that are broadly shared with all organisms, whose origins are billions of years old. These tend to encode very basic bits of biochemistry. There are genes that we share with all animals, or all mammals, or all primates, or all great apes. Genetic genealogies closely resemble evolutionary family trees, but not perfectly. This is largely because evolutionary family trees aren't shaped like trees. After only a few generations back they become matted webs, as ancestors occupy more than one position on your pedigree. Here's an extreme example from our prehistory: the lineages of *Homo sapiens* and *Homo neanderthalensis* split some 600,000 years ago. Both evolved independently for all that time apart, until 50,000 years ago when *Homo sapiens* turned up in their lands, and we all had sex. We know this because we've sequenced the Neanderthal genome, and if you are European, you have DNA that is clearly from them, and was introduced at that time. Within a few thousand years, they were gone, but their DNA remains alive in us. Some of that

Neanderthal DNA subtly influences the biology of living Europeans, including skin and hair pigmentation, height, sleeping patterns, and even a predisposition for smoking, even though that vice wouldn't be invented for a few hundred millennia.* In terms of an evolutionary tree, therefore, this introgression of Neanderthal DNA into you, if indeed you are of European descent, represents a loop. Trees don't have loops. Though genes are mostly passed down family trees, the trees themselves can be messy, and genes can enter a lineage from other directions, from ancestral cousins, or even, as we have seen, from a virus. They can also be lost in time just via the normal process by which genes get shuffled every time an egg or sperm is made.

Despite the messiness of ancestry, we can legitimately compare DNA in us, the Denisovans, Neanderthals and the other great apes, and try to infer whether the differences we see are significant.

HACNS1 isn't actually a gene.† It's a stretch of 546 letters of DNA called an "enhancer," sixteen of which are specifically different from what chimpanzees have. It's not a gene because it doesn't encode a protein, but what enhancers (or other bits of non-coding DNA) do is act as regulators for genes. Every cell with a nucleus contains every gene, but not every cell needs every gene to be active

---

* It just so happens that a naturally occurring genetic variant involved in something unrelated has an impact on the way we metabolize the chemicals in tobacco.

† The name stands for "human-accelerated conserved non-coding sequence 1." Human-accelerated because the individual changes in the sequence appear to have been acquired very quickly, which also might indicate its significance in having a function that is specific to us. Genes are generally and historically considered to be DNA that encodes a protein. It's not a watertight definition though, as other genetic elements are emerging, especially bits of DNA that make RNA, which has a function on its own without being translated into proteins. Anyway, you're probably getting the impression that evolution and biology are messy subjects with rules that mostly work most of the time, sometimes, with lots of exceptions. This is wholly accurate. Physicists never have this problem, the lucky devils.

at any one time. Enhancers tend to sit near the beginning of genes and act as instructions for that gene to be activated. Generally, we read sentences sequentially, from beginning to end and, in English, left to right. Genes are dotted all over the genome and can be read in any direction, in any order, on any chromosome, because unlike a book, they are never written in one go, or designed with a plan. A gene on chromosome 1 might activate a gene on chromosome 22. Enhancers and other regulatory bits of DNA control this apparent chaos.

We can test the function of an enhancer by looking at where and when it is active, and experimentally switching in mice embryos between the chimpanzee version and the human version. HACNS1 is active in a lot of tissues, including the brain, but it is buzzing with activity in the developing forelimbs, particularly in the tips of the bud that will develop into the paw. The same experiment with the chimp version of HACNS1 didn't show great activity in the same place. There's a similar pattern in the hindlimb buds too. As this chunk of DNA is an enhancer and not a gene itself, increased activity in the hands and feet is indicative of its role in turning on other genes, which are likely to be different in the hands and feet. Dexterity in human hands is essential for the tool crafting that we can do more skillfully than the other great apes, notably in the ability to rotate the thumb (which is longer in us relative to our other four fingers). Conversely, lack of dexterity and shortened digits in the feet was essential to our becoming bipedal. It's a striking theory that the rapid evolution of this short bit of DNA has had a significant role in altering the morphology of our hands and feet in ways that have become distinctly and uniquely human.

I could list a few more genes here that are intriguing clues to the genetic basis of uniquely human characteristics, and lots more will be discovered soon enough. Genes involved in brain development are particularly intriguing, because we have big, interesting brains.

Then again, because we have big, interesting brains, a huge number of genes are involved in the growth and maintenance of our neural matter. Some will promote the growth of new neurons, others the flourishing of connections between neurons. Some will be active in specific areas of the brain, especially in the neocortex, where so much of our insight and personality are centered. Many of these candidates will do many of these things and more, because evolution tinkers, and adapting and reusing something that already exists is easier and more efficient than inventing something from scratch.

Individual genes are often fascinating in their own right—though plenty are rather boring—and it is important that alongside the other 20,000 genes that every human bears, we continue to work out what they do, how they evolved, how they interact with the rest of our biology, and what happens when they go wrong. We also have to look at how they work with each other in the context of a functioning body.

Trippingly on the Tongue

There is one gene that is worth scrutinizing in much more depth. It's a gene that has much to say about our history and speaks volumes about evolution, and how we talk about evolution, and that is because it is a gene essential for speech. The story begins in Great Ormond Street Hospital in London in the 1990s. A family, known simply as KE, were being treated for a particular type of rare verbal apraxia, meaning that many members of the family had significant difficulty in turning sounds into syllables, syllables into words, and words into sentences. Fifteen people across three generations had these symptoms, most obviously the children, who would say things like "bu" instead of "blue," and "boon" instead of "spoon," among other verbal flubs. Further investigation showed that affected members of the family also had troubles that were not just related to articulation of words, but with more basic but specific movements of the face and mouth. When the same condition is seen in multiple generations in one family, we draw a pedigree and label the members who bear it. We can therefore assume that the random shuffling of genomes that happens when sperm and egg are made has not diluted the disease-causing DNA out of the lineage, but has been retained in those individuals. The inheritance pattern in the KE family pointed toward a single genetic defect being the cause. Though things are hugely more complicated now, at that time in the history of clinical genetics, most of the diseases that had been identified were indeed rooted in a single gene—conditions such as cystic fibrosis, Huntingdon's disease or hemophilia. In those ancient days of genetics, researchers would use a pedigree like this to hunt down a gene, and in 1998 Simon Fisher and his team found the sole cause of this family's speech and language problems. It was a gene that was named *FOXP2*, and since then has become an icon in genetics and evolution.

The gene *FOXP2* encodes a transcription factor.* These are proteins whose only function is to clamp onto very specific bits of DNA (such as the enhancer HACNS1 previously described). That way, one gene can control the activity of a second, a third, and so on, and a cascade of complex activity is triggered that helps to specify the different cells and tissue in a developing embryo. All genes are important, but some are more important than others, and transcription factors fall into that latter category. During your time as an embryo *in utero*, you grew from one single cell into trillions, carefully arranged into different types of cell, in different tissues doing very specific things. Transcription factors have a major role in the growth of an embryo. They function as controllers, busying away like foremen, setting up major building works, such as determining which end of an amorphous blob of cells is going to be the head and which the tail. Once that is in place, other transcription factors can set up ever more precise plans that specify "a brain goes up at this end," "in the brain area, eyes go here," "in the eye area, the retina goes here," "in the retina, the photoreceptors go here" and "among photoreceptors,

---

* A brief, necessarily tedious note on how we write about genes. A gene encodes a protein; both generally have the same designation, but genes are written in italics. *FOXP2* the gene encodes the protein FOXP2. Also, human genes tend to be written in capital letters, where the equivalent mouse gene is in lower case, but follows the same logic: *Foxp2* encodes Foxp2.

these ones are going to be rods." The details get more and more specific as the embryo develops, and the tissues differentiate into their mature fates. *FOXP2* is one of those which operates in the middle of those grand schemes of a developing embryo, and primarily has the effect of instructing the growth of more cells. When we look at where it is active in an embryo, it's in discrete areas all over the brain, clearly directing all sorts of neuronal growth, including in motor circuitry, the basal ganglia, thalamus and cerebellum.

Of the weapons in the arsenal of geneticists, seeing where the gene is active is just one. We can also extract the protein and see what it interacts with, a sort of molecular fishing trip. When we fish with *FOXP2*, it is fairly promiscuous, but some of its targets again offer alluring clues, such as a short stretch of DNA known as CNTNAP2, which is itself associated with speech disorders.

With all this in place, we have a gene that, when defective, causes a litany of speech and language disorders, and is active in various bits of tissue that are closely associated with speech. Other animals communicate orally, but in terms of sophistication, our language trumps even the closest by miles by every measure.* Given that we are the only organism that speaks with complex syntax and grammar, a genetic basis for our language skills is of great use in trying to demarcate ourselves as different from the other animals.

FOXP2 was not created *de novo* in us. In fact, it is an extremely ancient gene, as transcription factors often are. Similar versions are found in mammals, reptiles, fish and birds, many of which vocalize in some form. We know that in songbirds, their version of *FOXP2* is active in the brain when they are learning new songs from other males to woo females.

---

* With the possible exception of frequency: some animal communications are far higher or lower in frequency, and thus not audible to us, such as elephants, as mentioned on page 173.

In chimpanzees, their *FOXP2* is only two amino acids different out of the 700 that make up the protein, but the consequences are clearly significant—we speak and they do not. In Neanderthals, it is the same as in us, but other sections of their DNA may regulate differences in what the gene is doing. In mice, with whom we last shared a common ancestor about nine million years before the dinosaurs were wiped out, they have a version of *FOXP2* that is only four amino acids different. When we look at where the mouse *Foxp2* is active, it's in entirely equivalent places in the brain during development. When one copy of the gene is experimentally removed in mice, they display some abnormalities, one of which is a reduction in the number of ultrasonic peeps that pups usually make (if both copies are knocked out, the baby mice die after twenty-one days).

The fact that it is clearly essential for human speech and grammar, that it is different in us from the mice and chimp versions, and that it has undergone positive selection in *Homo sapiens* shows the elemental importance of *FOXP2*. It shows that this one particular gene is terribly important, but not all-important.

We can dissect the body at a number of different scales, and genetics is ultra-micro-anatomy. If we zoom out, the next useful resolution might be actual anatomy. After all, genes code the proteins that direct the cells that are assembled into our bodies. Anatomy changes over time: embryology is the study of how a single fertilized egg grows into an embryo, and developmental genetics is the study of the genes that moderate that growth. We think often only about adult vocal tracts, but it hardly needs stating that children are born immature, and this is relevant to understanding the development of speech. Tongues are large versatile muscles that aren't just the bit in your mouth laden with taste buds. They're rooted all the way down the larynx, and heavily innervated to control the movement and sensation that we need. In a newborn the tongue is almost entirely held within the mouth, which is so that the airflow of the larynx is connected to the nose, and the baby can breathe while nursing. As children grow up, the tongue descends into the larynx, and this enables the formation of full vowel sounds, such as "i" and "u."

There's a very important horseshoe-shaped bone in our throats called the hyoid. It sits under the chin, the horns pointing backward, and moves up and down when we swallow. It's intricately carved to accommodate twelve different muscle attachments, which gives us an idea of what a sophisticated piece of bone it is. Birds, mammals and reptiles all have versions of hyoids, but ours are much more intricate than all others, which is a reflection of the complex anatomical architecture required to create the vast range of sounds that comes so naturally to us, in combination with fine motor control of the muscles of the larynx and face. We think Neanderthals also had similarly elaborate hyoids, at least based on one specimen found in the Kebara Cave in Israel. Their overall anatomy was different from ours, not by much but enough for us to speculate that their hyoid would have been doing slightly different things to ours. But none of this is enough to think that Neanderthals couldn't speak; they had similar genetics, neuroscience and anatomy. That, for now, is the best we can do.

FOXP2 is significant in human evolution, but also in the evolution of science. It was one of the first genes to be characterized as causing a specific neurological defect when broken and has been singled out for that reason as having a significant impact on our nature, legitimately more than many other genes. It has been the subject of some breathless commentary as "*the* language gene," and indeed as *the* trigger that fired the gun of our modernity. We will come to the role of speech in our behavior in a few pages, but it's key to understand that complexities of genetics in relation to anatomy and behaviors are both inscrutably complex and poorly understood. We can see that *FOXP2* is essential, but it is active in a whole bunch of cells in the brain, and therefore has influence over other biological functions. The KE family's troubles were not restricted to speech. They also struggled with lexical tasks, where the subject has to distinguish between real words and nonsense words that obey general rules of English, such as "glev" or "slint." That is a psycholinguistic effect. Again, this indicates the complex interplay between our motor and cognitive skills.

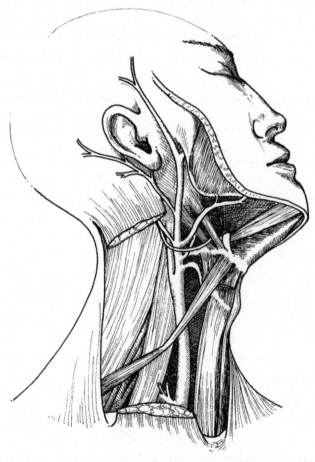

A most intricate hyoid

The great linguist Noam Chomsky has suggested a romantic notion that there was a switch, a spark that lit the fire of language in us, where the best any other creature could manage was grunts and gestures. His timescale is plausible, thousands of generations, but it implies a focused linearity founded in a singular trigger.

Evolution does not work like that. Modern genetics has shown that humans have been much more mobile than previously thought, and have interbred continuously in and out of Africa, facts that don't lend support to a linear view of our deep history. Furthermore, speech is not one single thing. The physical capability of speech, with its anatomy and neural control of that anatomy, is not distinct from the neural control of speech. We are a system, made up of small interconnected cogs and parts. We have to consider how brains develop, and what genes are doing in that process. Neural tissue is highly specialized, and comprises hundreds of different cell types, each with their own genetically determined identity. Cells become neural tissue, and once on that pathway, grow, migrate and adorn themselves with synapses and dendrites that connect with adjacent cells or ones that are millimeters or centimeters apart (which is a long way if you're a neuron). After you were born, your brain went through a process of synaptic pruning for many years until your teens, when the connections between neurons were cut back or reinforced in order to streamline thinking and learning. All of that is controlled by genes and their interaction with our environment. The point is that one gene involved in this crazily complex building project is likely to have multiple effects on different tissues, and dozens if not hundreds of genes are going to play a role.

Speech is the audible output that rests upon dozens of highly complex, interconnected biological phenomena. *FOXP2* is necessary but not sufficient. A highly structured hyoid is necessary but not sufficient. A neurological framework with the ability to fine-tune motor control of the muscular fibers in the larynx, tongue, jaw and mouth, as well as forming a psychological basis capable of perception, abstraction and description is absolutely necessary, but not sufficient. And of course, when we speak, we disturb particles of air, which vibrate the drums of our ears and trigger the similarly complex process of hearing. Without ears or air, there is no speech. Genes are templates, brains are frameworks, the environment is a canvas. We separate out each of these parts only in order to understand the bigger picture, but let us not pretend that they all popped into being at once.

A far better way of understanding the acquisition of speech, and indeed the acquisition of any emergent characteristic in humans, is the selection and genetic-drift model, and via a changing interaction between culture and our genes, a mutation in *FOXP2* set in a framework from which language could develop. We don't know if the Neanderthals had that same framework; we can reasonably imagine that they did, given their similarities in material culture, morphology, and a version of *FOXP2* that is the same in us and different from chimps. I suspect that they were speakers, but it will take a very clever experiment to help clarify that question, one of which I cannot quite conceive, at least not yet.

Speak Now

When it comes to the origin of language, the trouble is that talking leaves no fossils.

The biology of speech is complex enough, but there is much more to it than simply the capability, as we have seen. Complex communication is essential to what we call behavioral modernity, that is, how we are today, compared to how we were when we were not like how we are today. We will come to that in a few pages.

We are biologically programmed for speech. We have the neurological, genetic and anatomical template that greenlights the possibility of language. We have a latent ability to acquire language, by copying the sounds of the people around us. Some birds have that too: they learn their love songs from each other. Each bird species has a few songs, enough that a well-trained ear can identify a species by its sound, though many have regional dialects (as indeed some whales do). In contrast, humans currently speak over 6,000 distinct languages, all of which are continually evolving, most of which are heading for extinction, and you probably know tens of thousands of words and can deploy them at will. We also learn syntax and grammar from those around us, our brains a software platform specific to language acquisition. Anyone with children will have heard them make endearing errors in grammar because they have, without instruction, generalized a rule. My four-year-old daughter says "swimmed" as the past participle of "swim," because her brain has learned the rule that actions from the past are typically denoted by adding "-ed" to a word. We have to learn exceptions to rules, while having an inherent ability to transpose one grammatical rule to another word. That is a colossally powerful software platform.

Words change meaning over time too. Cromulent words are continually added to embiggen our lexicon to great behoovement or crinkum-crankum; others end up in a linguistic midden. Angry grammar pedants fantasize that language always degrades away from some imagined inviolate form, without recognizing that words and language are in perpetual evolution through usage, and that the original meaning of a word isn't necessarily the same as the current usage. Linguists nobly and successfully try to draw evolutionary trees for words and for languages, which is far harder than evolutionary biology because the spoken word does not echo through time like a bone turned to rock. Nevertheless, we construct historical relationships between languages, and build evolutionary language trees. They can be broadly informative: a hypothetical Proto-Indo-European language gave rise to a European trunk that spawned Slavic branches, Germanic boughs and Romance boles, and from an Indo-Iranian trunk, Iranian, Anatolian and hundreds of other languages and dialects. These types of tree don't tend to account for the continuous horizontal transfer of words looted from other languages as people moved around the world.

In the previous paragraph, I used English words borrowed, derived or adapted from Hindi, Anglo-Saxon, Norse, Latin and *The Simpsons.** English enjoyed a massive invasion of words from

---

* Loot: लूट, meaning "steal"; tree: from Old English *trēow*; gave: from the Old Norse *gefa*; nobly: from Latin *nobilis*, meaning "high-born"; embiggen: a cromulent word meaning "to enlarge," from the season 7 episode "Lisa the Iconoclast" (1996).

1066 when William conquered; Brits took on a continuous stream of Old Norse words during the years that the Vikings bothered the British coast; the Romans brought Latin with them—our incredibly rich language represents a gestalt mish-mash that reflects our history, both genetic and cultural. As genetics gets more sophisticated in mapping historical migration, we're even seeing surprising interactions between who we are and what we speak. It seems that the indigenous people of Vanuatu were completely replaced around 400 BCE with another population from the Oceanian Bismarck Archipelago, but retained the same language over that transition. In this extreme example, cultural transmission of a language is completely disassociated from genes.

Symbolism in Words

All those words and meanings that you store in your brain, and all the words that you are yet to learn, are not simply in a look-up table to be accessed when you need them. You understand words. If you look at a nose, you recognize that what you are looking at is a nose, because through experience, you know what noses look like. If you read the word "nose," you are not looking at a nose. Yet you also know what I'm talking about. On top of that, I could add other words, adjectives to enhance the idea; if you think of a humungous red nose you've linked together three independent concepts—size, color and an object—and fused them not only as a symbolic description of an imagined object, but an abstract one that is not based in reality, but of which you still can conceive. The plasticity of symbolism is complex and clever.

With the exception of onomatopoeia, linguists generally think that the symbolism in words is arbitrary. "Buzz" does sound like its meaning, but *deux, zwei, ni, tše pedi, rua, núnpa* and *tsvey** all mean an ordinal number above one and less than three, yet there is no inherent reason why each of those words has come to mean the same thing.

Consider the sperm whale, improbably summoned into existence above the planet Magrathea in *The Hitchhiker's Guide to the*

---

* French, German, Japanese, Sotho, Maori, Lakota Sioux and Yiddish.

Galaxy. Surprised by his genesis, he cheerfully pondered the origin of words as he fell:

> And wow! Hey! What's this thing suddenly coming toward me very fast? Very very fast. So big and flat and round, it needs a big wide sounding name like . . . ow . . . ound . . . round . . . ground! That's it! That's a good name—ground!
>
> I wonder if it will be friends with me?

Poor old whale. Paradoxically, he had a fulsome vocabulary to compare words when sourcing a *de novo* one for the lethal land beneath him. The implication is that there is an inherent property of the word "ground" that relates to its physicality. A study in 2016 suggested that there is the faintest wisp of inherence to certain words, and that this is universal. Linguists looked at a hundred words that qualify as basic vocabulary, from 62 percent of the world's languages. These words included pronouns, basic verbs of motion, and nouns for body parts and natural phenomena, such as "you" and "we," "swim" and "walk," "nose" and "blood," "mountain" and "cloud." The analysis was probabilistic, meaning that they used statistics to calculate the possibility that sounds in words in unrelated languages are similar at a higher frequency than by chance alone. The English word that we use to describe the visual perception of electromagnetic spectrum energy at a wavelength between 620 and 750 nanometers is "red." In other European languages that are closely related in time and space, red words also contain a prominent "r" sound: *rouge, rosso, røt.* But that "r" sound is also more likely than by chance a key part of the words for "red" in languages unrelated to Indo-European ones. The word to describe the protruding part of the centre of our faces with two holes in it, primarily used for detecting odor, is one that across the world is likely to have a nasal or "n" sound in it.

That doesn't necessarily suggest that words with similar sounds have a common root, but it may indicate that the neurological framework that enables speech identifies a very basic underlying grammar that makes some words gravitate toward certain sounds. Our brains may gently steer us toward certain sounds somehow resembling the thing they describe.

Even with that in mind, this effect is subtle for unrelated languages, and took some deep data analysis to spot. Symbolism in words is not generally inherent. Words across the world for a nose might have a tendency to sound nasal, but *nez, Nase, hana, nko, ihu, p̃hasú* and *noz* are not a nose, and they only describe a nose via consensus.

And so, any language must be predicated on the ability to attribute one thing with another. With those tens of thousands of words you know, you can order them and construct a learned syntax to convey meaning, and you do so every time you speak without making a total gallimaufry of it. Isn't that clever? I looked up "gallimaufry" just to find an unusual or archaic word that was unknown to me, but even without knowing it, you can figure out exactly what it means from the context in that sentence.

A word is a symbolic unit of meaning to represent a thing, or an action, or an emotion. But when a parrot parrots, we don't think that it is applying symbolism to the sounds it makes. It is merely copying. We also communicate non-verbally, via symbolic gestures, in the sense that the gesture itself doesn't necessarily imitate the action that follows. Some of our gestures are demonstrative of the required action, such as a typical repeated beckoning action of a finger or hand that says "there to here." Others are clearly not, and meaning is agreed in culture. A raised hand, flat palm facing away, means either "stop" or "hello" in many cultures. This wave is demonstrated by the naked man on the gold-aluminum plaques aboard the spaceships Pioneer 10 and 11, in case they ever find alien life as they speed across the Milky Way; I always thought that was a bit weird, as that hand action might mean "I wish to open-palm punch you in the face" or even "please violently impregnate me and then decimate my species" to any handed alien that is unaware of convention so arbitrary it can mean opposite things to many humans.

This concern is borne out in understanding the non-verbal symbolic gesture of chimpanzees and bonobos. A hold to the top of the arm from a bonobo may mean "climb on me," and in chimps "stop what you're doing," especially to a young one. A big scratch on the upper forearms might mean "initiate grooming" to a bonobo, or "travel with me" to a chimpanzee. A raised arm might mean "I'm going to climb on you" to a bonobo, but "get that thing" to a chimp.

Typically, a great many of the bonobo gestures mean "initiate copulation" or "initiate GG rubbing" (see page 100), the most obvious one being a legs-akimbo presentation of their genitals, which seems to straightforwardly mean "interested?" Let's hope that the aliens who find the Pioneer spaceships are not as horny as the bonobos. Chimpanzees are not quite so groinally obsessed, but even with that comparative chastity, waving a branch, or a touch on the shoulder seems to mean "let's get it on" in both species of *Pan*. We can reasonably say that the gestures are symbolic and learned, not just because they don't necessarily resemble the requested action (though presenting one's genitals has a fairly obvious meaning), but because the meanings are different in two different species.

We also know now that other mammals are capable of learned vocal symbolism. Prairie dogs and vervet monkeys have alarm calls that are specific to the predator, and they act accordingly. For the monkeys, a low grunt warns of an eagle above, and in response, the monkeys look up and hide under trees; a hoo-haah pant follows a leopard being spied, and the monkeys head up for the thinnest branches of trees that will support their weight, but not a leopard's; a high-pitched shriek warns of a snake, and the correct response is to stand up on two feet and survey the ground.

Audible symbolism is not just limited to primates either. Stridulation is the vigorous rubbing of two body parts that produces the sound of crickets at night, and a thousand other insects, as they advertise their sexual availability. It isn't merely saying, "I am here and up for it," as many vary the tone to mark territories or as alarms, as well as for sex. And while we're on insects, the famous waggle dance of honey bees is nothing but a symbolic gesture, not audible, but indicating information about distance and direction to water or juicy nectar.

That animals communicate is not at all surprising. So far, our exploration of animal communication has revealed that the ability of non-human animals to pass on information via explicit messaging or via symbolic gestures is widespread and common. All the available evidence so far also suggests that it is nothing like ours, at least in terms of the number of units of meaning that they have in their vocabulary. As I've said elsewhere in this book, it is worth noting that almost all nature goes unobserved by us, and we ought to display some humility for the things we haven't yet discovered. We've known about infrasonic vocalization in elephants since the mid-1980s, whereby they communicate with other elephants using frequencies well below our audible range, which has the advantage of travelling many miles with little distortion. We're beginning to get a good idea of how dolphins and some whales convert air vibrations into aquatic noise; in these two cetaceans, it may have some similarities with our own larynxes, but in other types of whale, such as the baleen family, we don't really know.

In captivity, many great apes have been taught a vocabulary of arbitrary symbolic gestures by instruction from their scientist keepers. Some primate celebrities such as the bonobo Kanzi, who was born at Georgia State University in 1980, or the gorilla Koko, born at San Francisco Zoo in 1971 (and who died in June 2018), have managed hundreds of signs as a basic language. Whether these are simply learned by rote, or they have some understanding of the signs themselves is debatable. A dog will go nuts at the sound of the word "walk" or "park," not because it knows what a nice green space is, but simply because of the repeated association between that word and a jolly jaunt. My wife and I used to use the French word *glace*, so as not to alert our young children to a potential ice cream treat. But like dogs with no knowledge of French, they soon worked out that when we said *glace* in the context of a sentence and in the park, an ice cream often followed.

These captive apes have a significantly high number of signs in their vocabulary, hundreds, which is about the same as a three-year-old. But the non-human apes lack any sense of grammar, or ability to generate a sentence—a typical three-year-old can manage to generate a five-word sentence with ease—*I really want ice cream*. None of the other apes demonstrates structure in their communications, or tense. What children do easily is fundamentally different. Genes, brains, anatomy and environment provide the canvas on which children learn complex, abstract, arbitrary and symbolic words, grammar, syntax and language, and they do it without even trying.

Verbal, or at least audible, symbolism is not limited to humans, nor is gestural symbolism. As with other examples in this book, we must be wary of implying that similar behaviors in animals and us have a shared evolutionary origin. The genetics of *FOXP2* in us and other vocalizing animals shows that there is a clear evolutionary precedent for the genetics, neuroscience and anatomical mechanics of making noise with our mouths, from birds to monkeys to dolphins to us (and this isn't shared by insects, who make noise with their limbs and other body parts). The application of meaning to those symbols—noise and gesture—looks to be particular to certain species, but we are leagues ahead in range and sophistication.

We need to speak, and we need to describe, and we need to abstract, and we need to predict and exchange information about our thoughts and the thoughts of other people. Maybe in the wild, away from our prying minds, gorilla communications are much more sophisticated, by a mechanism that we can't yet see. Their communication has evolved to suit what gorillas do, and not as evolutionary and neurological templates for understanding how we do what we do. For now, language is unique to us.

Unique to us, though probably Neanderthals were like this as well. And we will probably discover that the Denisovan people were also potential speakers, if we ever find any more of their mortal remains.

Symbolism Beyond Words

We speak, with all that software and hardware in place. There was no toggle switch that flipped us from being like the other apes into what we are now. We think that full language capabilities must have been in place by around 70,000 years ago, because that is when the diaspora from Africa happened, and all the dispersed populations in that exodus have sophisticated language. If we are right and the Neanderthals and Denisovans had sophisticated language too, then we can consider one of two options available to us: either language was in place before we three humans split, more than 600,000 years ago, or the physical capability for sophisticated language was in place in us and them, and we began to speak independently.

Whichever way speech and language emerged in humans, it was a transition, with all those necessary but not sufficient pieces being nudged in one way or the other, by chance, by selection. The fact that it was a transition, not a revolution, means it took time. But we don't have a very good handle on how long this took. The separation of our lineage from that of the other great apes occurred six or seven million years ago. We know it was definitely after that. Our brains became significantly larger from about 2.4 million years ago and continued to grow, so it was definitely after that too, as we don't think a small brain has enough firepower for fully operational speech and language. *Homo sapiens* comes into being from 300,000 years ago, according to specimens from Morocco and east Africa, and by 100,000 years ago we have bodies pretty much the same as we do today.

Forty thousand years ago, we have art. This is huge step up in our grasp of symbolism. At that time, humans all over the planet were beginning to display what scientists sometimes call "the full package," that is, behavioral modernity. On a giant southern isthmus on the island of Sulewesi, part of Indonesia, there are caverns that were the homes of people over thousands of years. About eight paces from the entrance of one particular cave is a mural, consisting of 1.5 meters of drawings. There are twelve hands—in fact, the shadows of hands because they have been stenciled—with red ochre blown through a thin tube to outline the hands of a long-gone person. Nearby, there is a drawing of a fat pig, and a "pig-deer" called a babirusa. These were drawn something like 35,000 years ago, and the oldest of the handprints is 39,000 years old.

As of October 2018, the record for the oldest known figurative art by *Homo sapiens* is claimed by the former residents of Borneo, just to the west of Sulewesi. The Lubang Jeriji Saléh caves are remote, and it is a treacherous journey to visit them. But they are cavernous, cathedral-like spaces, their walls adorned with a gallery of thousands of animals, people and hand stencils that spans tens of thousands of years. A large image of a banteng, a type of local wild cattle, is painted in ochre on a low ceiling deep inside the cave. The artist must have crouched, head tilted far back, or lay on her or his back, like Michelangelo painting the ceiling of the Sistine Chapel. Using the slow but measurable decay of radioactive uranium in a fine crust of calcite taken from the tail end, Australian researchers think that this cow was painted a minimum of 40,000 years ago, and a maximum of 52,000.

Over in Europe, at around the same time, people were creating art in very similar ways. Southern France is littered with caves adorned with pictures of astonishing beauty and skill that date from around this time all the way into the near present. Lascaux, near Montignac, is probably the most famous, a Pleistocene art gallery from a much more recent 17,000 years ago, displaying more than 6,000 figures, interpretations of hunts, with horses and bison, felines, the extinct colossal elk *Megaloceros giganteus*, and abstract symbols whose meaning we can never understand. People painted in charcoal and hematite and dabbed them onto the walls as pigments in suspensions with animal fats and clay. They are breathtaking.

To the west, the Chauvet-Pont-d'Arc Cave has the oldest wall art in Europe, again with beasts in relief, from hunts, and hunters—cave lions, hyenas, bears and panthers, *oh my!* The oldest of these were painted 37,000 years ago, according to the most up-to-date studies in 2016.

And then there is the *Löwenmensch*—the Lion Man of the Hohlenstein-Stadel. In the hills between Nuremberg and Munich in Swabian Germany there are caves that have yielded one of the most important works ever crafted by an unknown artist. Around 40,000 years ago, a woman or man sat somewhere in or near that cave, with the detritus of a hunt scattered around. They took a piece of ivory, a tusk from a woolly mammoth, and carefully considered that it might be the right material, shape and size for something that they had been pondering. Now extinct, cave lions were fierce predators at that time, posing a threat to people, and also to the animals that people would hunt and eat. That person thought about the lions, and how formidable they are, and maybe wondered what it would be like to have the power of a lion in the body of a human. Maybe this tribe revered the cave lions out of fear and awe. Whatever the reason, this artist took that mammoth ivory, a flint knife, and patiently carved the tusk into a mythical figure.

It is a chimera, a fantastic beast that is made up of the parts of multiple animals. Chimeras exist throughout all human cultures for most of history, from mermaids, fawns or centaurs, to the glorious monkey-man god Hanuman, to the Japanese snake-woman nure-onna, to the Wolpertinger, an absurd and mischievous Bavarian part-duck part-squirrel part-rabbit with antlers and vampire teeth. Today, we have reached the ultimate manifestation of a 40,000-year interest in hybrid creatures in genetic engineering, where elements from one animal are transposed into another, and hence we have cats that glow in the dark with the genes of deep-sea crystal jellyfish *Aquorea victoria*, and goats that produce dragline silk from the golden orb weaver spider in their udders.

The first of these that we know of is the *Löwenmensch*. It is an extraordinary piece of work, around twelve inches tall, the figure of

***The Lion Man of
Hohlenstein-Stadel***

a man with a lion's head, and an important piece in understanding our evolution. Of the artist, it shows profound skill, fine motor control and foresight in selecting the right bone and having a plan to carve this figure. It shows an understanding of nature, and reverence of the animals in the ecosystem that impact upon the lives of those people. Crucially it shows a willingness to imagine a thing that does not exist.

The figure is a man, as determined by the genitals, and has seven stripes cut into the left arm, almost like tattoos. It was found deep inside the cave at Hohlenstein-Stadel in 1939, in an almost secret vault, a kind of cubbyhole that also held other objects—carved antlers, pendants and beads. The inference is that these objects were precious, possibly things with totemic significance. Nearby in the Vogelherd Cave, figures of a mammoth and a wild horse have been found, and the ornately carved head of a cave lion. Perhaps cave lions at this time were iconic of a ceremonial cult, and the cut marks on the arm meant something important on this mythical creature. Perhaps.

A few miles to the east, we find the earliest example of another charm—the Venus of Hohle Fels (see illustration on page 5). There are many examples of prehistoric sculpted female bodies. They are generically called Venus figurines, after the first one discovered in the Dordogne in the 1860s by Paul Hurault, the eighth Marquis de Vibraye, who, noting the pronounced incision representing the vulva, called it *Vénus Impudique*—the "immodest Venus." The Venus of Hohle Fels is the most ancient of these figures, probably again around 40,000 years old. It is the oldest depiction of the human body.

This Venus is also an abstraction. It is clearly a human body, but a heavily distorted one, with features that are well beyond realism. The breasts are colossal, and the head is tiny. She has a huge waist, and engorged labia. These enhanced sexual characteristics are also seen in some of the other Paleolithic Venus figurines, which has led to speculation that these were fertility charms, or even goddesses of fertility. Some people have suggested that they might be pornography. While there is no shortage of art by men depicting sexualized women, we cannot know the motivation of the Venus sculptor. The similarities between the few Venus figurines that remain do suggest a sexual dimension to their existence, and imagining that they are fertility amulets is no more or less speculative than considering that they are the fantasy of a Paleolithic artist. We're not sure why the heads are often small: it might be to do with perspective, that you can't actually see your own head, so relatively from one's own vision it is small, and looking down, breasts may look disproportionally larger; though that doesn't account for the fact that the artist could've seen the heads and bodies of other people. Maybe it was an artistic choice. If in one million years' time, you discovered a Francis Bacon portrait or the Bayeux tapestry isolated out of any context, you might have questions about what was on the minds of those artists. We will never know what the Paleolithic sculptors were thinking. What we do know is that their minds were not different to our own.

There are flutes or recorders from this time in Germany as well, hollow tubes with finger holes carved from the bones of mute swans, mammoths and a griffon vulture. Percussion or drumming instruments may well have preceded these, as hitting things to make rhythmic noises does not require the same cognitive imagination as crafting a multi-toned whistle with fingering (with apologies to the drummers of the world).

There is some dispute about the precision of these dates. The techniques used to date rocks and the art upon them are not always agreed upon, and the margins of error can be thousands of years. For the broader sweep of human evolution, the precise dates aren't pivotally important. By 40,000 years ago, there are clear, unequivocal depictions of figurative art in multiple forms, and clear evidence for imagination, abstract thought, music and profound fine motor skills. Something had changed.

The geographical spread is important, not just in itself, but because Borneo is a long way from Europe. The art we have found in caves in Europe is from around the same time, which means one of two things: the skills to create such works were shared by an ancestor common both to the Southeast Asian artists and the European ones, which means tens of thousands of years earlier. Or it means that people in Borneo started drawing independently, at around the same time. Because of the paucity of artistic remains in the geological record, the parsimonious explanation is the second. For the idea of a common artistic ancestor to be justified, we would need to see much older art, spread geographically from Europe to Indonesia.

All these artifacts show clear signs of the hallmarks of modernity. These artists had "the full package." They had a rich culture, and a reverence for their environment, which implies an emotional recognition of their position in nature, and in their own tribes. They thought about sex, and imagined dreamlike beings that cannot exist, but somehow told them things about their lives. This behavior would be spread across the world in the next ten or twenty thousand years, though not necessarily from a single root. Evidence for fuller packages is seen in Siberia, northeast Asia, southeast Asia and Australia in the millennia that followed, though we mustn't presume that these people learned their new cognitive skills from a direct lineage; they may have evolved on their own in those places. However it emerged globally, those first artists had music, painting and wore fashions. They were us.

Until 2018, we thought we alone were like them. In northern Spain, there are caverns set into the Cantabrian coast, and deep in one known as El Castillo, there are large squares, like an eighteen-inch frame, in red and black paint. Inside one frame is the outline of the back legs of an animal, which could be bovine, but it's impossible to know for sure. In another panel is the image of the head of an animal, again possibly a bison or maybe a horse. There are also linear signs, geometric shapes and a weird image almost resembling a figure, which is oddly reminiscent of Picasso's 1955 silhouette portrait of Don Quixote.

These paintings and two other examples of Spanish cave art were dated in early 2018, and in all cases (according to one analysis), they appeared to be older than 64,000 years. The only people in Europe at this time were not *Homo sapiens*. They were *Homo neanderthalensis*. Neanderthals were in a small but absolute way the ancestors of most Europeans today via inter-species breeding. They were here in Europe first, hundreds of thousands of years before our direct ancestors had begun to trickle out of Africa. If these dates are correct, those Neanderthal people were thinking about the hunt and painting their prey on walls a good 20,000 years before we would invade their territory.

The earliest examples of figurative art were not done by us, but by our cousins. We've known for a while that Neanderthals had culture, and earlier we discussed the potential of their vocal capabilities. The caves that sit beneath the Rock of Gibraltar have been a rich source of Neanderthal activity, and have revealed their culture, diets and one example of something akin to art. In Gorham's Cave, there is a series of scratches that looks like the remains of a big game of noughts and crosses. The marks are very deliberate, one groove being carved by the repeated action of more than fifty strokes around 40,000 years ago. The scientists in Gibraltar who manage this amazing site have tried to emulate its creation, and rule out these marks as being a by-product of butchering meat, or tailoring skins. These lines were carved deliberately and for no obvious reason.

We can go back further. For us, there have been a few clear examples of modern behavior tens of thousands of years before the Neanderthals left this world, and we became the last humans. Neanderthals were never in Africa as far as we know. The Blombos Cave in South Africa overlooks the Indian Ocean and has been a hotbed of evidence for modern human lifestyles, from more than 70,000 years ago, including bone tools, specialized hunting, the use of aquatic resources, long-distance trade, beaded shells, use of pigment, and art and decoration, notably in ochre shales, carefully engraved with geometric criss-cross patterns. Nearby in the caves at Pinnacle Point we find micro-engineered quartzite blades and red-ochre pigments made for an unknown purpose. The date of these antiques is around 165,000 years ago. And even further back, there are fossil freshwater mussel shells from Trinil in Java that have been engraved with inch-long grooves in sharpened peaks, a sort of bivalve doodle. The age range for these is a bit flabby, but they were carved sometime between 380,000 and 640,000 years ago. That predates any other evidence of intentional and non-utilitarian craftwork. The only people on Java at that time were our deep evolutionary cousins *Homo erectus*.

There are plenty of traces of modern skills and behavior long before the so-called "cognitive revolution" 45,000 years ago. But they are sporadic blips in time, and not permanent, as the evidence vanishes from the continuous archaeological record. This material culture has become permanent by 40,000 years ago, give or take a few millennia. By then, the Neanderthals are gone. By 20,000 years ago, we have it all: art, jewelry, tattooing kits, weapons including spears, boomerangs and barbed harpoons, and it is all over the world.

If Only You Could See What
I've Seen with Your Eyes

Art, craft and culture require a sophisticated mind. They require language too, to communicate the complexity of those abstract creations and their meaning to our families and wider social group. We cannot know the order by which we acquired these traits, and it might be unhelpful to even think of this evolution in a step-by-step way. The changes are slow, gradual and subtle to get all the pieces in place for who we are today.

We can think about the progress of language acquisition as a child might, which is different from the evolutionary process because the framework required is already in place in a child. Nevertheless, first you name objects—*cave lion*—and later, you attach action to named objects—*approaching cave lion*. Next you can associate more detailed and useful attributes—*two large cave lions approaching.* In a social group, conveying this type of information is essential, just as the calls of a vervet monkey alerting its pals to an eagle are. You are conscious of this situation, and it is useful to know that someone else is—*are you aware of the two large cave lions approaching?*—because they can further impart useful details that will save you from wasting precious resources—*the two approaching cave lions are full cos they just ate Steve.*

To imagine the mind of another is key to our cognitive development, and language has to be part of that too, because we have to transfer complex information between individuals and groups. When babies are born, they almost immediately have the ability to recognize faces, most frequently those of their mother and father. Eye contact comes naturally to infant humans. We can test how long their eyes fall upon an object or person, and infer what they are more interested in. Babies prefer open eyes, and over the months of development will recognize different emotions in the faces of others—joy, anger, sadness, fear, disgust. They will begin to express their own emotional state in their faces and voices too, which go from simply pooling pain, hunger, tiredness and fear into one category—"this feels wrong"—to the full gamut of human emotions, hopefully at some point in their life. We know that a few animals can read human faces, and maybe even limited emotional states of those humans. Sheep are very good at identifying individual humans. Experiments in 2017 showed they could be trained easily to recognize specific faces—including Barack Obama's—though shepherds have known this for a while.* We saw earlier that those very smart Caledonian crows learned faces that were a threat and ones that were benign, and could remember this information for years. Dogs, as any owner will know, seem pretty good at recognizing the emotional state of their human, and in tests, will change their facial expression much more if they know a human is looking at them.

The ability to assess the emotional state of another is mind reading. You are trying to understand what it is that another mind wants or needs. That's limited if you're just using non-verbal cues.

---

* Though this experiment seemed silly, sheep are very good model animals for terrible neurodegenerative diseases such as Huntington's. In some of these types of brain disorders, neurons die and specific functions are lost, including the ability to recognize people's faces.

It also limits the communication to the present, which is something that humans do not do. Of course, beasts think into the future, and recall the past. They think about feeding and reproduction, and the success of their offspring. Birds and other animals, including the squirrel, think forward in time by squirreling food away for another day, and then have to recall where they put their nuts. Many salmon return to the precise place of their birth, even though they have spent most of their life in the ocean.

These memory feats are not the same as in us. We are extreme mental time-travelers. We think about the past, and not just in a perfunctory or rote-learned way. Here I am thinking about Steve, my 40,000-year-old human. It's not so difficult to imagine his thought process when he encountered the cave lion who spelled his demise—ours would be much the same today. But I can also try to imagine what that person was thinking when they sat and carved the *Löwenmensch*, or one of those bosomy Venus statues. And we can think about the future. Not just what the next meal is going to be, but make plans for my mum's birthday in July, or what my next book will be. I like thinking about what songs I want played at my own funeral, and hope that the guests will enjoy them.

Leaping forward and backward in time enables our innate ability to recognize the mind of another conscious being. Consciousness is a poorly defined concept, and means many things to many people, including a sense of self, sentience, an ability to experience or to feel, and other things. Much has been made of the question of whether animals have consciousness or not, but it really depends on what you mean by consciousness. Clearly animals are sentient and experience their environment. Many animals can recognize themselves, and can engage with the mind of another creature within or outside of their own species. Do they have an ineffable inner life? Will we be able to establish a neurological basis of our own consciousness and then compare it to that of other animals?

These are all outstanding questions for much more research, and another book.

For now, we can recognize consciousness in another human, even though it is poorly defined, and we often think we see the same in other animals, regardless of whether it is true or not. In fact, we are so sensitive to another consciousness that we imagine it everywhere. Humans are so keyed into seeing faces as representations of a mind that in our fiction we give personalities to animals that fall well outside of any meaningful definition of consciousness—insects, tardigrades, crabs. Pareidolia is the psychological phenomenon of seeing faces in inanimate objects—Jesus in a piece of toast, a face on the surface of Mars. Our brains know faces are important, so they recognize the pattern of a face even if there can be no mind behind it. We are also so plugged into other consciousnesses that we detect agency when there is none. It's extremely useful to attribute agency to dangerous situations and adapt one's behavior accordingly. An animal might do this by many means: many mammals are innately repelled by chemical clues in the urine of a predatory fox or coyote; birds are fooled by scarecrows. We're smarter than birds, but don't have the noses of rabbits, so we largely rely on visual and auditory cues. Stumbling across the freshly mutilated body of Steve, it pays to think *that looks like the work of a cave lion, I must flee!* rather than simply acknowledging *Steve's not looking great right now.*

Poor old Steve. The result of a mind so highly attuned to others is that like with faces, we attribute a mind to mindless events. A creak in a floorboard as the house cools at night and the wood shrinks is creepy because our brains are instantly trying to detect agency behind this noise, rather than rationally processing the thermodynamics of the situation. I am reluctant to delve too deeply into this, because it is an area only of speculation and not particularly scientific, but it is attractive to think that this might be a significant part of the explanation for the existence of religion. Our minds seek agency from another conscious mind, rather than dumb nature, either living or inanimate. This is a force powerful enough for us to imagine ghosts; it could conceivably also be the genesis of gods.

Mercifully, the full package of our evolution has also equipped us with the ability to override this cognitive short-circuit and seek the real reason why things without obvious agency happen. However we made the gods, with careful thought we can also tuck them away again.

Know Thyself

Another part of this complete cognitive package is not just knowing others, but knowing yourself. Recognition that you are an individual with agency and self-determination. The mirror test is a standard of ethology nowadays. Can you recognize that the image reflected from a mirror is not a moving picture or someone mimicking your actions, but is actually you? It's designed to test the ability of an organism to be visually self-aware. In some versions of the test, you spot a bit of dye on the forehead of the participant, without them knowing, and see if they try to touch their own heads where the dot is. This way, they are recognizing that the mark on the individual in the mirror is in reality on their own head. By the time human children are about two years old, they will direct their hands to the dot on their head. If you have a baby, this is a fun and easy experiment to do from six months old.

A few animals have also passed this test, to much acclaim. Bottlenose dolphins and killer whales appear to pass, sea lions do not. Three elephants have been tested by placing a red cross on their heads, not visible without a mirror, and of those, only one, called Happy, acknowledged and repeatedly tried to touch the spot with her trunk.* Of the super-smart birds, so far only a single magpie

---

* As a control, an odorless transparent cross is also painted on the elephant's head, which Happy completely ignored.

has shown any ability to recognize that the reflection is their own body.

I wonder how significant mirrors are in the grand scheme of cognitive evolution. Certainly, this test shows a level of thinking that relates an abstraction to reality—"that is me, but it is also not actually me"—but it's a strange sort of test from which to extrapolate huge inferences. It tests visual recognition, when many organisms don't primarily rely on sight for sensory input. Surely a dog would be better off doing some sort of smell-mirror test? Also, it's testing an artifice. Animals presumably can see and detect parts of their own body, in the full absence of mirrors in their lived experience. Does that make them somehow quantitatively less self-aware than us? I don't think it does. Gorillas don't pass, though maybe ones in captivity with a lot of human familiarity might. Then again, eye contact is generally a sign of extreme aggression in gorillas, so maybe getting them to spend time staring at an image of a gorilla is not reflective of their cognitive capabilities. In 1980, the psychologist B. F. Skinner also challenged the significance of the mirror test by intensively training pigeons to pass it. Bribed with food, the pigeons were shown dots that they first had to twist their heads to see, and second, they could see in the mirror. After a few days' training, the pigeons would identify spots on their own bodies, by only looking at the mirror. They had learned to pass the mirror test, for a handful of seeds.

I'm not saying that the mirror test is invalid; it's more that being self-aware is certainly a facet of a high-cognition mind, but there are many ways to be self-aware other than clocking oneself in a looking glass. It's quite an anthropocentric test, as it relies on the assumption that being able to see oneself in a mirror is an important symptom of a state of mind. Toads spend a lot of time sitting very still after reversing into damp holes, but we don't consider that fortitude of endurance to be a neuroscientific benchmark of any sort, even though it's clearly important to the toad. We talk of the traditional five senses, but in reality, there are many more than that. Proprioception is a significant one here—awareness of one's own body in space; another is interoception—an awareness of one's internal bodily state: try sitting perfectly still (like a toad) and counting your heartbeats by nothing other than feeling them in your body. These are also key expressions of a sense that you are a body in space, independent of the environment.

Self-awareness is essential for recognizing that you are a being which is separate from everything else. It is part of the conscious experience of being human, and the experience of being in some other animals.

Je Ne Regrette Rien

Within conscious experience, we endure and enjoy psycho-physiological states that are a hallmark of the human condition. Or feelings, as normal people call them. It's tempting to attribute emotions to other animals. Our pets sometimes look joyous and happy, or listless and miserable. One of our cats, Moxie, is just a horrible person—grim, aloof, sour, and uninterested in any contact with me, effectively nothing more than her derisible butler. Our other cat, Looshkin, is more like a dog, with boundless enthusiasm and a generally happy, affectionate and slightly crazed outlook. But look at all the heaps of anthropomorphizing I pile upon them both. In fact, I have no idea what they're thinking about their inner experience, or their emotional state. We cannot know what it is like to be another animal, cat, bat or human. We make the mistake of assuming that their experience is like our own, and their emotional states are reflected in the same way that ours are.

Darwin was very interested in this in the nineteenth century, and expanded his thoughts into a whole book on the subject in 1871. Since then animal behaviorists over the years have tried to understand emotions in animals, and attempted to rationalize them. One strategy has been to separate basic emotions from more complex ones—happiness, sadness, disgust and fear are all straightforward, visceral emotions, whereas jealousy, contempt and regret are more complex and cerebral. Grief or mourning has been observed in many primates and some elephants, with heartbreaking descriptions of gorillas holding wakes, or Gana, an eleven-year-old gorilla in Münster Zoo in Germany, who in 2008 became famous after newspapers printed pictures of her carrying the lifeless body of her infant child.

It takes an unnecessarily hard scientific heart to not acknowledge these examples as anecdotal evidence for complex emotional states in animals. But we are still ultimately hampered by simply not being able to ask them what they feel, or for them to volunteer their complex emotions to us. We are however in the age of neuroscientific techniques in which we can try to read brains better and thus make more scientific inferences about the internal emotional status of an animal. With these new techniques we are beginning to find out whether their experiences are like ours. This is a neonate area, but one example is worth exploring.

The French singer Édith Piaf may well have regretted *rien*, but most of us will experience it *beaucoup*. Regret is such a specific and complex emotion, to feel disappointment for a decision that was, with the clarity of hindsight, suboptimal. Many people express disdain for regret, like Piaf, on the stubborn grounds that it is futile to admonish oneself for past actions. Lady Macbeth paraphrases another French idiom, this time from the fourteenth century,* when she declares:

> Things without all remedy
> Should be without regard: what's done, is done

Admirable though this sentiment is, things didn't really work out well for the Macbeths. Others have suggested that you should only regret the things you haven't done, rather than those that you have.

---

* *Mez quant ja est la chose fecte, ne peut pas bien estre desfecte.* Translation: "But when a thing is done already, it cannot be undone."

While that sounds high-minded, it's not really very practical, and is the preserve of glib motivational quotations. I'm more in line with Katharine Hepburn:

> I have many regrets, and I'm sure everyone does. The stupid things you do, you regret . . . if you have any sense, and if you don't regret them, maybe you're stupid.

Regret is an explicitly negative emotion: to feel disappointment for the way things could have been, if only you had acted differently in the past; to feel sadness or anxiety at having failed at something, or having made a poor decision. There is a morality naturally built into regret, that you both could have and should have behaved differently. "It seemed like a good idea at the time." I love that phrase, for capturing the essence of regret, from the short-term and trivial—"one more glass of wine is just the thing before I head home"—to matters of permanence and consequence.

The complexity of conscious thought required for this sense to be felt is rich. You need two aspects of mental time travel. First, a perception of the past, a recognition that there were multiple options at that time, and an ability to conceive of imaginary outcomes dependent on an alternative version of events. You also need an ability to imagine a different future. Ultimately, the function of regret is not to wallow in your errors, but to learn from them as an expression of free will: "Next time, I'll do it differently, and the benefits will be greater, or at least, less bad." We do it all the time. As an emotion, its existence rests on so many very human qualities. And it turns out rats express regret too.

Again, we must be stringently careful not to assume that behaviors in animals that resemble things familiar to us are the same. Violent and coercive sexual intercourse in animals is not rape, though as discussed earlier, in some cases the comparisons are striking, at least in some dolphins and sea otters. Until we can ask an animal what it is feeling and thinking, we have to make do with rigorous scrutiny, and hold ourselves back from assuming that they are feeling what we do in similar situations, especially when what we experience is complex. A well-designed experiment can certainly help though.

Restaurant Row is one of those experiments. Designed by psychologists Adam Steiner and David Redish from the University of Minnesota, it is an octagonal arena with four dining areas in opposite corners. It's a bit like a food court in a shopping mall, with multiple restaurants serving different food styles. There are four meals available to the rats in Restaurant Row: banana, chocolate, cherry and plain. Rats, like us, really don't enjoy waiting for food, and each flavor is only made available to a rat after a random-length wait. There's also a beep, which descends in pitch to indicate how long they have to wait for the food—the higher the initial tone, the longer the wait. Rats enter the arena and are trained to recognize the tone and the associated wait, and the flavor of the reward that follows.

In the experiment, each of the rats is known to have a natural preference for one of the flavors over the other three. The experiment is to show the rat a long delay for the flavor they like the best, but give them the opportunity to switch to another flavor, rather than hold out for their preference. Say you've got a cherry-loving rat, and she knows that she has to wait twenty seconds for her favorite flavor. But it's a long wait, and the rat bails out after fifteen seconds. She cuts her losses in the hope that she will get a banana-flavored snack in the meantime. But the banana wait turns out to be another twelve seconds, which means that in total she has waited twenty-seven seconds and got some food that she's just not that into. She gambled by not being patient, and lost. It's like being hungry in the mall, and you really want some sushi. But you're impatient, and the queue for the sushi bar is long because sushi takes time to prepare. So you hedge your bets and switch to a pizza because the queue is shorter, only to see a big batch of sushi arrive as soon as you've committed to the pizza queue. The sushi sells out. You're not that into pizza, and you instantly regret that decision.

The rats regret their change of mind too. How can we tell? They look at the flavor they prefer but didn't get. To say they looked longingly would be crossing the line into anthropomorphic assumption, but they do turn their heads and stare. In some situations, they waited a shorter time to get a less-favored meal—the pizza was ready quicker, even though you really wanted sushi, but you eat it anyway. What you might experience here is disappointment, rather than regret. When merely disappointed, they didn't turn their heads.

More importantly though, the next time they're faced with that same gamble, they wait. They have recognized that an impatient decision was punished, and learned to play the odds more conservatively.

If this still seems like interpreting a very ratty behavior as a direct analogue to complex human emotions, Steiner and Redish looked at what was happening in their diners' brains while they were being subjected to these scenarios. The orbito-frontal cortex (OFC) is an area of our brains where neurons are known to be turned on when we experience regret. Experiments have been done where human volunteers are subjected to gambling choices which have been secretly contrived by scientists. After their bets are made, and lost, they are shown what they could've won if they'd made a different choice; thus, the scientists have rigged the experiment so that they can induce regret in the participants. People with damage to this part of the brain don't experience it, and report no regret in response to negative consequences after poor decisions. You can't ask a rat to report how it's feeling. Instead, the rats in Restaurant Row had their OFC monitored for excitement during their meal choices. Specific cells came alive when thinking about each of the flavors, including their favorites. The same cells sparked when they had passed up on their favorite flavor, had a longer wait, and turned back to gaze upon the missed opportunity. Cherry-loving rats were still thinking about cherry when they gambled and got banana.

While this all sounds a bit cute, understanding the neural correlates in rats of complex human emotions has clinical potential. Some psychiatric conditions include an absence of regret or remorse, or feelings that follow, such as anxiety, which normally might contribute to making a different or better decision in the future. Understanding the circuitry that is damaged or misfiring is where we start to fix it.

The fact that a similar brain area is alert when expressing regret in two distantly related mammals might suggest that the mechanism being adopted to feel this emotion is ancient. Rats and us are separated by tens of millions of years of evolution, but this result doesn't mean that every species between them and us also expresses regret in a similar way—we just don't know. Other animals need to be tested in similar ways. Until then, if regret is the emotion in us that induces a change in behavior when faced with the same situation in the future, we can at least be certain that these rats regret.

Teach a Village to Fish . . .

We've seen that there is little physical difference between a woman or man 100,000 years ago, and you or I today. We can see almost certainly that language is older than the onset of the full human package. Our brains are not significantly different from when we were just dabbling with art, and indeed they don't seem fundamentally different from those artists who were not us, but our cousins the Neanderthals. The symptoms of modernity have been with us for tens of thousands of years longer than its arrival. Evidence is found scattered in Europe and Indonesia by 40,000 years ago. There are examples of modernity in Africa and Australia within a few millennia after it's seen in Europe as well. These make a genetic basis for the change unlikely, as they are spread over the world with no interaction, no gene flow, between these peoples. If we are to assume that all the humans that spread around the world had emerged from Africa, and were genetically similar, then it's unlikely that they independently will have enjoyed the same DNA mutations that triggered the arrival of a complex mind. If the Paleolithic people of the world were biologically similar, the question is this: why did it take so long to become modern when we were physically ready for thousands of years?

There are many pieces of this puzzle that remain elusive. These are areas of research that are beginning to blossom, such as theory of mind, and the nature of consciousness. They are questions that have languished in fascinating philosophical realms for decades and centuries, and are beginning to be examined with the more precise scientific tools of the twenty-first century. We inch toward a better understanding of those areas as they become entwined with neuroscience.

There's one idea that I think is crucial, that has been emerging in the last few years, but is not yet discussed widely, though I hope it will be soon. It is that population size and structure changed, and with those changes, modernity followed. The full package came about because of how we organized our society.

The first clue to this theory is that populations seem to grow larger at the time of the onset of modernity, in multiple locations. We see it in Africa 40,000 years ago, and at a separate time in Australia, more like 20,000 years ago. These expansions may be in line with the local environment, simply that as the climate changed, life became easier. They might also be a manifestation of our huge migrations. No other creature has moved permanently in such a short period of time—within 20,000 years of leaving Africa, we had settled in Australia.

We also see the opposite effect: a loss of cultural sophistication in societies whose populations do not grow, migrate or are cut off from a bigger populace. For example, Tasmania became an island around 10,000 years ago, as the last Ice Age thawed and the seas rose, and was separated from mainland Australia by what Europeans named the Bass Straits. The indigenous people of Tasmania managed to maintain a tool kit of only twenty-four pieces in that isolation, and lost the skills to make dozens of others over thousands of years in the Neolithic. Indigenous Australians on the mainland developed more than 120 new tools during the same period, including multitoothed bone harpoons.

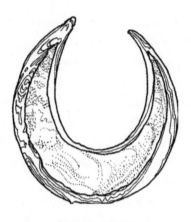

A Javanese fish hook

In the Tasmanian archaeological record, we see the gradual disappearance of fine bone tools, the loss of the ability to make cold-weather clothing, and perhaps most significantly, the degradation of fishing technology. Hooks and spears for catching cartilaginous fish vanish from archaeology, as does evidence of fish bones (although they did continue to forage and eat crustaceans and sessile mollusks). When Europeans arrived in the seventeenth century, the indigenous people expressed both surprise and disgust at the colonizers' skill at catching and eating large fish, yet 5,000 years earlier it was a key and thriving part of their diet and culture.

Scientists who are interested in the full package have developed models to try to understand how the cultural transmission of skills is affected by the size and structure of a population.* In this way, they can test how and why we see the hallmarks of modern behavior come and go, and eventually stay in the archaeological record. These are effectively equations that model how an idea or skill is

---

* Notably spearheaded by Mark Thomas and colleagues at University College London, and Joseph Henrichs at Harvard University, among others.

passed around in a community. They plug in hypothetical numbers for the size and density of a population, and a skill level for an imagined expert task—maybe knapping an arrowhead, or tooting a flute—and then they run simulations that work out how that skill level can be transferred between people. Mathematical models of this sort are pretty technical, but what they are effectively doing is saying: "Here are people with a very particular set of skills, which can be taught to others. How does the size of a population affect the efficiency of teaching?"

The answer appears to be "enormously." Larger populations enable the transfer of complex cultural skills with far greater efficiency than smaller ones. The maintenance of skill levels is heavily dependent on population size (which is also affected by migration). According to the models, small populations, especially isolated ones, will lose skills through an inefficient transmission. When populations grow, they accumulate culture more readily. Only we do this. Though there is a smattering of other examples of cultural transmission in other animals, we do it all the time.

I don't think that demographics is necessarily an obvious link to how we became who we are, which might be why it has been relatively neglected. But when we look at what humans are, it makes perfect sense. We are social, meaning that we depend upon the interactions with others for our own well-being. We are cultural transmitters, meaning that we pass on a wealth of knowledge that is not encoded in our DNA. We do this horizontally, not just vertically, meaning that we teach to people who are not our children, but are our peers, and may not even be genetically close kin. And we are highly skilled and creative, but that expertise is not distributed evenly throughout our populace—some people have skills that others don't, and when we need to find out how to do something, we ask an expert.

There's a second reason why this might not be as popular an idea as I think it ought to be. For many years in the infancy of evolutionary biology, scientists fiercely debated a question absolutely fundamental to Darwin's greatest thought, natural selection: *what is being selected?*

Of the potential answers available, they scale all the way from the gene, to the individual, to the family, to the larger group, to the species. We put all that to bed in the middle of the twentieth century with unequivocal evidence that the answer is the gene. A gene encodes a phenotype—that is, the physical manifestation of a piece of DNA—and differences in those physical manifestations in a population are visible to nature as a means of selecting what works better. The gene that encodes that phenotype is what is transmitted from generation to generation, the unit of inheritance. A gene for processing goat's milk after weaning was selected in humans over a gene that did not permit digestion of a nutritious drink. Individuals are merely carriers of genes, which drive the necessity of procreating simply so that the existence of the gene is perpetuated.

This gene-centric view of evolution was discovered and developed by some of the titans of twentieth-century biology—Bill Hamilton, George Gaylord Simpson, Bob Trivers, and others—and was immortalized in one of the great works of popular science, Richard Dawkins's *The Selfish Gene*. It is correct, and now textbook. What this new model suggests is that there is selection for cultural transmission of things that are adaptive and therefore of great benefit to us, pivoting not on genes but on a population. We biologists are rightly groomed away from ideas about group selection because they are not correct—the data does not fit the idea that evolution acts on groups. But cultural transmission is not encoded in DNA, and in some ways, is exempt from the precise mechanisms that happen in the formation of egg and sperm that engineers genetic difference in a population, and therefore is subject to Darwinian evolution.

Put together like that, it seems obvious that the demographic structure of a society is going to be essential in maximizing the way information and skills are transmitted within a group. Any group of people relies on an internal organization to be effective. It seems from these models, that our modernity—the full package of being the humans that we are today—is dependent on us being able to accumulate culture, to pass it on, and to pass it on in a society that grew to optimize the overall success of its members.

This is a territory that is actively being researched right now. It is the model that I think is right, for what that is worth, though much more work is needed. A tiny proportion of the ground has been dug to uncover our pasts. A fraction of the genes of our ancestors have been sampled. As ever in science, the answers are never complete, and we mould and carve ideas, throw them away if the data doesn't fit, or build them up if it does. The idea that demography was an essential pivot in the ascent of us is an idea that is young.

The truly amazing thing is that Darwin was thinking along the very same lines, one and a half centuries ago. He writes in *The Descent of Man*:

> As man advances in civilization, and small tribes are united into larger communities, the simplest reason would tell each individual that he ought to extend his social instincts and sympathies to all members of the same nation, though personally unknown to him. This point being once reached, there is only an artificial barrier to prevent his sympathies extending to the men of all nations and races.

The Paragon of Animals

I've written quite a lot of these words in an Italian café near where I live. It's early on a Friday evening right now, and it's buzzing. I'm the slightly odd man sitting alone with his fourth coffee and a heap of books. It strikes me that restaurants are wonderful places for observing the full package of human evolution. There's a school nearby and there are teachers and pupils in here. It's very family friendly, and there's a baby being cooed at by someone I guess is a grandparent, but might be unrelated. People fork farmed food cooked by fire into their incredibly complex mouths using forged metal tools. A couple on a date may well have more fun to come later tonight. The manager oversees the chefs in the kitchen, who interact with the waiting staff, who interact with the customers. And everyone is talking.

Next time you are in a café, take a moment to watch what is really happening. Every transaction is an exchange of information. All those dynamics are the outcome of a biological and cultural evolution unique to this ape. We exhibit sexual preferences and activities that are diverse and by choice, yet comparable to behaviors seen in other animals. We have separated sex from reproduction with a boundary that is rarely breached. We've taken technology to levels of sophistication indistinguishable from magic.

Our brains have grown and funded these abilities and behaviors that sometimes differ by degree, sometimes by kind, even if they may look much the same. Our minds have expanded beyond our brains, at least metaphorically, because humans are a social creature that transmits ideas through time as well as space, and very few animals do that as effectively. Where we stand apart most significantly is in cultural accumulation and transmission. Many animals learn. Only humans teach.

There is cultural transmission of ideas in a few other species: tool use in females of a pod of tech-savvy dolphins in Australia; maybe the knowledge of who is scary and who is not to a Caledonian crow. These are few and far between, and we will discover more examples in time. Humans do it all the time and have done for millions of years. Due to the nature of my work, I stand in front of thousands of people every year and tell them stuff that I have learned. I am related to almost none of them. We accumulate knowledge, and pass it on. That is what this book is, what all books are.

Here is a secret: I did none of the actual research that I have written about. I've never been to Indonesia and seen the hand stencils of our forebears; I've never sat in the Senegalese savannah looking at chimps patrolling a wildfire. I've never been to Shark Bay and witnessed dolphin cows wearing sponges on their beaks. I hope that one day I will. Some of you have, and scientists have, and they did this to satisfy their curiosity, and by proxy, ours. They wrote those things down and applied the accumulated knowledge of 10,000 years to check if it was right, and shared those ideas with others, also to check, so that humans would learn something that they didn't know before. I read their work, all of the scientific papers referenced at the back of this book, and used my experience of teaching and learning to crunch those ideas, and try to synthesize them into something new, and gestalt. I wrote them down, and my editors and a couple of scientists used their skill and experience to question my words and ideas, and knock them into shape to make it all easier for others to understand. The designers and typesetters put it together, and Alice Roberts drew on her knowledge and skill with pen and ink to draw some beautiful pictures. And all together, we made this thing you are holding, for no other reason than to share some ideas.

Every journey of every human is built upon thousands of years of accumulated knowledge, founded upon billions of years of evolution. Our culture is part of our biological evolution, and it is wrong to try to separate them. Our minds evolved because it was advantageous and appropriate to do so, and the selection of our cognitive faculties and our minds are only important in the context in which they evolved. Mutations in our genes provided a physiological change that set the template for speech, and the processing power to allow that speech to develop into complex communication. That helped elevate our thought processes so that a mind with consciousness akin to our own today could be built out of the necessity of anticipating another mind's thoughts. None of this happened in a flash, there was no spark, there was no one thing that set off this chain. Our minds evolved, and as we know, evolution is slow and messy and convoluted. A mind that can time travel and read others, speech, dexterity, fashion, the joy of sex, all are the result of a lumpy continuum, emergent properties underwritten by that force of evolution.

A living organism is an integrated system. Though that seemingly catastrophic fusion of two chromosomes actually set up the unlikely framework of the human genome, there is no single genetic change that made us *Homo sapiens*. Take a machine such as a car: it didn't become a car with the addition of the gearbox, or the steering wheel, or any other single part. All of the parts are together what makes it a car, some essential, some less so, but none definitional. You can lose a limb in an accident, or even possess an extra chromosome, and you are still a human. We are so much more complicated than motor vehicles, even though we have roughly the same number of genes as an average car has parts. More and more we are finding that genes do many things. There isn't a gene for speech, though *FOXP2* is clearly essential. There isn't a gene for creativity, imagination, spear throwing, dexterity, consciousness, or even cultural transmission. There wasn't a moment when we were not *Homo sapiens* before, and suddenly we were after one gene mutated. Our genomes are unique to us, and provide the evolved framework on which humanness could emerge.

In Christian cultures, we talk of the Fall, where humankind became sullied by shaking off the shackles of our creation. I don't care for that story much. If anything, we fell upwards, slowly and incrementally, and away from the thoughtless brutality of nature. The Lord knows there's plenty of wickedness in humans, and though we mostly reject the primal urges that we might have inherited from four billion years of indifferent evolution, the numbers are on the side of Hamlet's angels. We almost never murder, we almost never rape, we create and teach all the time, and learn almost at the same rate.

The picture of how we came to be is only going to get more complicated as we continue to discover. I suspect that soon we will find more contemporary species of human who lived alongside us within the last 300,000 years, and that we will find more humans who bred with us in that time too. We should revel in this complexity and celebrate the fact that we alone are capable of understanding it.

Evolution is blind and evolutionary progress is a misnomer; natural selection carves and winnows according to the ever-changing status quo. Just like all living things, we struggle for existence, but we also try to ease the struggles of others.

We must, however, acknowledge, as it seems to me, that man with all his noble qualities, with sympathy which feels for the most debased, with benevolence which extends not only to other men but to the humblest living creature, with his god-like intellect which has penetrated into the movements and constitution of the solar system.

Charles Robert Darwin wrote those words in 1871. He is my hero, for better or worse, and though he was so very right about some of the most important ideas that anyone ever had, like all scientists, he was wrong about others. He was right about the evolutionary pathway of humans, and simultaneously he was woefully wrong about the evolution of women, whom he thought were intellectually inferior to men. At least, part of his incomparable legacy is that we now know this to be incorrect.

Nevertheless, with the use of the word "man" to mean "human," Darwin draws *The Descent of Man* to a conclusion, writing, "with all these exalted powers—Man still bears in his bodily frame the indelible stamp of his lowly origin."

Our genes and our bodies are not fundamentally different from those of our nearest cousins, ancestors or even our deep relatives. As for lowly origins, that is a matter of judgment. We are evolution's creatures, forged, carved and etched out of forces beyond our control, just like every living thing is. With those forces behind us, we took evolution's work, and by teaching, we created ourselves, an animal that together became more than the sum of its parts.

Remember the alien naturalist come to Earth to study us. In Carl Sagan's novel *Contact*, a real fictional alien intelligence does scrutinize humankind—they've been watching us for thousands of years. In that story, we send a scientist according to their instructions, and upon meeting her, the alien speaks:

You're an interesting species. An interesting mix. You're capable of such beautiful dreams, and such horrible nightmares. You feel so lost, so cut off, so alone, only you are not. See, in all our searching, the only thing we've found that makes the emptiness bearable, is each other.

Life is continuous on Earth, endless forms most beautiful. We force discrete classifications upon that continuum to help us make sense of a planet bursting with life through eons. You sit somewhere on that trajectory, unique in trying to figure out your place in all of this. There is no dedication at the front of this book. Instead, it is for you.

Sign your name below, and work backward:

You are _____
You are *Homo sapiens*
You are a great ape
You are simian
You are a primate
You are a mammal
You have a backbone
You are an animal
We are the paragon of animals.

ACKNOWLEDGMENTS

All of the following human people contributed in some way to the ideas I've written about in these pages, and I am very grateful to all of them, even the ones that don't exist: Alex Garland, Andrew Mueller, Aoife McLysaght, Beatrice Rutherford, Ben Garrod, Caroline Dodds Pennock, Cass Sheppard, Cat Hobaiter, the Celeriacs, David Spiegelhalter, Elspeth Merry Price, Francesca Stavrakopoulou, Hannah Fry, Helen Lewis, Henry Marsh, Ieuan Thomas, James Shapiro, Jennifer Raff, John Ottaway, Jon Payne, Kate Fox, Lynsey Mathew, Mark Thomas, Michelle Martin, Nathan Bateman, OAs Elite Coaching Crew, Rachel Harrison, Robbie Murray, Rufus Hound, Sarah Phelps, Simon Fisher, Stephen Keeler and Tom Piper. And to Georgia Rutherford, who is the very best of us.

Particular thanks go to the unreasonably talented Alice Roberts for her highly evolved hands, and for guiding mine. To Matthew Cobb, whose edits are a joy to behold. To Will Francis for our ongoing journey, and most of all to Jenny Lord and Holly Harley, two of the most thoughtful, fun and brilliant humans with whom to share ideas and craft stories.

REFERENCES

Adams, Douglas, *The Salmon of Doubt* (William Heinemann, 2002)

Aggrawal, Anil, "A new classification of necrophilia," *Journal of Forensic and Legal Medicine* 16(6): 316–20 (August 2009)

Aranguren, Biancamaria, et al., "Wooden tools and fire technology in the early Neanderthal site of Poggetti Vecchi (Italy)," *PNAS* 115(9): 2054–59 (February 27, 2018)

Aubert, M., et al., "Early dates for 'Neanderthal cave art' may be wrong," *Journal of Human Evolution* 125(12): 215–17 (December 2018)

———, "Palaeolithic cave art in Borneo," *Nature* 564: 254–57 (November 7, 2018)

———, "Pleistocene cave art from Sulawesi, Indonesia," *Nature* 514: 223–27 (October 8, 2014)

Bailey, Jeffrey A., et al., "Genome recent segmental duplications in the human," *Science* 297(5583): 1003–07 (August 9, 2002)

Berna, Francesco, et al., "Microstratigraphic evidence of in situ fire in the Acheulean strata of Wonderwerk Cave, Northern Cape province, South Africa," *PNAS* 109(20): E1215–E1220 (May 15, 2012)

The Incredibles, written and directed by Brad Bird, Pixar Studios, 2004

Blasi, Damián E., et al., "Sound-meaning association biases evidenced across thousands of languages," *PNAS* 113(39): 10818–23 (September 27, 2016)

Bonta, Mark, et al., "Intentional fire-spreading by 'firehawk' raptors in northern Australia," *Journal of Ethnobiology* 37(4): 700–18 (December 2017)

Brown, D. H., "Further observations on the pilot whale in captivity," *Zoologica* 47(1): 59–64

Cair, Osvaldo, "External measures of cognition," *Frontiers in Human Neuroscience* 5: 108 (October 4, 2011)

Camille, Nathalie, et al., "The involvement of the orbitofrontal cortex in the experience of regret," *Science* 304(5674): 1167–70 (May 21, 2004)

Ciani, Andrea Camperio, and Elena Pellizzari, "Fecundity of paternal and maternal non-parental female relatives of homosexual and heterosexual men," *PLoS ONE* 7(12): e51088 (December 5, 2012)

Chapela, Ignacio H., et al., "Evolutionary history of the symbiosis between fungus-growing ants and their fungi," *Science* 266(5191): 1691–94 (December 9, 1994)

Clayton, Nicola S., et al., "Can animals recall the past and plan for the future?," *Nature Reviews Neuroscience* 4: 685–91 (August 1, 2003)

Coe, Malcolm J., "'Necking' behavior in the giraffe," *Journal of Zoology* 151(3): 313–21 (March 1967)

Connor, R. C., et al., "Two levels of alliance formation among male bottlenose dolphins (*Tursiops* sp.)," *PNAS* 89(3): 987–90 (February 1, 1992)

Cornelis, G., et al., "Retroviral envelope *syncytin* capture in an ancestrally diverged mammalian clade for placentation in the primitive Afrotherian tenrecs," *PNAS* 111(41): e4332–E4341 (October 14, 2014)

Dannemann, M., and J. Kelso, "The contribution of Neanderthals to phenotypic variation in modern humans," *American Journal of Human Genetics* 101(4): 578–89 (October 5, 2017)

Darwin, Charles R., *The Descent of Man, and Selection in Relation to Sex* (John Murray, 1871)

D'Anastasio, R., et al., "Micro-biomechanics of the Kebara 2 hyoid and its implications for speech in Neanderthals," *PLoS ONE* 8(12): e82261 (December 18, 2013)

D'Souza, Gypsyamber, et al., "Differences in oral sexual behaviors by gender, age, and race explain observed differences in prevalence of oral human papillomavirus infection," *PLoS ONE* 9(1): e86023 (January 24, 2014)

Deaner, Robert O., et al., "Monkeys pay per view: Adaptive valuation of social images by rhesus macaques," *Current Biology* 15: 543–48 (March 29, 2005)

Deaner, Robert O., et al., "Overall brain size, and not encephalization quotient, best predicts cognitive ability across non-human primates," *Brain, Behaviour and Evolution* 70: 115–24 (May 18, 2007)

Deecke, Volker B., "Tool-use in the brown bear (*Ursus arctos*)," *Animal Cognition* 15(4): 725–30 (July 2012)

Dennis, Megan Y., et al., "Evolution of human-specific neural SRGAP2 genes by incomplete segmental duplication," *Cell* 149(4): 912–22 (May 11, 2012)

Dunn, Dale G., et al., "Evidence for infanticide in bottlenose dolphins of the Western North Atlantic," *Journal of Wildlife Diseases* 38(3): 505–10 (July 2002)

Emery, Nathan J., "Cognitive ornithology: The evolution of avian intelligence," *Philosophical Transactions of the Royal Society B* 361(1465): 23–43 (January 29, 2006)

Jaffe, Karin Enstam, and L. A. Isbell, "After the fire: Benefits of reduced ground cover for vervet monkeys (*Cercopithecus aethiops*)," *American Journal of Primatology* 71(3): 252–60 (March 2009)

Epstein, Robert, et al., "'Self-Awareness' in the pigeon," *Science* 212(4495): 695–96 (May 8, 1981)

Esnault, C., G. Cornelis, O. Heidmann and T. Heidmann, "Differential evolutionary fate of an ancestral primate endogenous retrovirus envelope gene, the EnvV *syncytin*, captured for a function in placentation," *PLoS Genetics* 9(3): e1003400 (March 28, 2013)

Fiddes, Ian T., et al., "Human-specific NOTCH2NL genes affect notch signalling and cortical neurogenesis," *Cell* 173(6): 1356–69. e22 (May 31, 2018)

Fisher, Simon E., and Sonja C. Vernes, "Genetics and the language sciences," *Annual Review of Linguistics* 1: 289–310 (January 2015)

Foster, Emma A., et al., "Adaptive prolonged postreproductive life span in killer whales," *Science* 337(6100): 1313 (September 14, 2012)

Fujita, Masaki, et al., "Advanced maritime adaptation in the western Pacific coastal region extends back to 35,000–30,000 years before present," *PNAS* 113(40): 11184–89 (October 2016)

Geßner, Cornelia, et al., "Male–female relatedness at specific SNP-linkage groups influences cryptic female choice in Chinook salmon (*Oncorhynchus tshawytscha*)," *Proceedings of the Royal Society B* 284(1859) (July 26, 2017)

Goodall, J., *The Chimpanzees of Gombe: Patterns of Behavior* (Belknap Press, 1986)

Graham, Kirsty E., et al., "Bonobo and chimpanzee gestures overlap extensively in meaning," *PLoS Biology* 16(2): e2004825 (February 27, 2018)

Grayson, Kristine L., et al., "Behavioral and physiological female responses to male sex ratio bias in a pond-breeding amphibian," *Frontiers in Zoology* 9(1): 24 (September 18, 2012)

Guerzoni, Daniele, and Aoife McLysaght, "*De novo* origins of human genes," *PLoS Genetics* 7(11): e1002381 (November 2011)

Gumert, Michael D., and Suchinda Malaivijitnond, "Long-tailed macaques select mass of stone tools according to food type," *Philosophical Transactions of the Royal Society B* 368(1630): 20120413 (October 17, 2013)

Han, Chang S., and Piotr G. Jablonski, "Male water striders attract predators to intimidate females into copulation," *Nature Communications* 1, article number 52 (August 10, 2010)

Harmand, Sonia, et al., "3.3-million-year-old stone tools from Lomekwi 3, West Turkana, Kenya," *Nature* 521 (7552): 310–15 (May 20, 2015)

Harris, Heather S., et al., "Lesions and behavior associated with forced copulation of juvenile Pacific harbor seals (*Phoca vitulina richardsi*) by southern sea otters (*Enhydra lutris nereis*)," *Aquatic Mammals* 36(4): 331–41 (November 29, 2010)

Hart, B. J., et al., "Cognitive behaviour in Asian elephants: Use and modification of branches for fly switching," *Animal Behaviour* 62(5): 839–47 (November 2001)

Henrich, Joseph, "Demography and cultural evolution: How adaptive cultural processes can produce maladaptive losses: the Tasmanian case," *American Antiquity* 69(2): 197–214 (April 2004)

Henshilwood, C. S., et al., "Emergence of modern human behavior: Middle Stone Age engravings from South Africa," *Science* 295(5558): 1278–80 (February 15, 2002)

Henshilwood, Christopher, et al., "Middle Stone Age shell beads from South Africa," *Science* 304(5669): 404 (April 16, 2004)

Higham, Thomas, et al., "Testing models for the beginnings of the Aurignacian and the advent of figurative art and music: The radiocarbon chronology of Geißenklösterle," *Journal of Human Evolution* 62(6): 664–76 (June 2012)

Hobaiter, Catherine, and Richard W. Byrne, "Able-bodied wild chimpanzees imitate a motor procedure used by a disabled individual to overcome handicap," *PLoS ONE* 5(8): e11959 (August 5, 2010)

Hoffmann, D. L., et al., "U-Th dating of carbonate crusts reveals Neandertal origin of Iberian cave art," *Science* 359(6378): 912–15 (February 23, 2018)

Ishiyama, S., and M. Brecht, "Neural correlates of ticklishness in the rat somatosensory cortex," *Science* 354(6313): 757–60 (November 11, 2016)

Joordens, Josephine C. A., "*Homo erectus* at Trinil on Java used shells for tool production and engraving," *Nature* 518: 228–31 (February 12, 2015)

Jónsson, Hákon, et al., "Speciation with gene flow in equids despite extensive chromosomal plasticity," *PNAS* 111(52): 18655–60 (December 30, 2014)

Kaminski, Juliane, et al., "Human attention affects facial expressions in domestic dogs," *Scientific Reports* 7: 12914 (October 19, 2017)

Kilpatrick, Dean G., et al., "Drug-facilitated, Incapacitated, and Forcible Rape: A National Study," National Crime Victims Research & Treatment Center report for the US Department of Justice (2007)

Krützen, Michael, et al., "Contrasting relatedness patterns in bottlenose dolphins (*Tursiops* sp.) with different alliance strategies," *Proceedings of the Royal Society B* 270(1514) (March 7, 2003)

Krützen, Michael, et al., "Cultural transmission of tool use in bottlenose dolphins," *PNAS* 102(25): 8939–43 (June 21, 2005)

Lahr, M. Mirazón, et al., "Inter-group violence among early Holocene hunter-gatherers of West Turkana, Kenya," *Nature* 529: 394–98 (January 21, 2016)

Larson, Greger, et al., "Worldwide phylogeography of wild boar reveals multiple centers of pig domestication," *Science* 307(5715): 1618–21 (March 11, 2005)

Linden, David J., *The Compass of Pleasure: How Our Brains Make Fatty Foods, Orgasm, Exercise, Marijuana, Generosity, Vodka, Learning, and Gambling Feel So Good* (Penguin, 2011)

Lipson, Mark, et al., "Population turnover in remote Oceania shortly after initial settlement," *Current Biology* 28(7): 1157–65 (April 7, 2018)

Marean, C. W., et al., "Early human use of marine resources and pigment in South Africa during the Middle Pleistocene," *Nature* 449: 905–08 (October 18, 2007)

McBrearty, S., and A. S. Brooks, "The revolution that wasn't: A new interpretation of the origin of modern human behavior," *Journal of Human Evolution* 39(5): 453–63 (November 2000)

McLysaght, Aoife, and Laurence D. Hurst, "Open questions in the study of *de novo* genes: What, how and why," *Nature Reviews Genetics*, 17: 567–78 (July 25, 2016)

Mitani, John C., et al., "Lethal intergroup aggression leads to territorial expansion in wild chimpanzees," *Current Biology* 20(12): R507–R508 (June 22, 2010)

Nair, Smita, et al., "Vocalizations of wild Asian elephants (*Elephas maximus*): Structural classification and social context," *Journal of the Acoustical Society of America* 126(5): 2768 (November 2009)

Neill, James, *The Origins and Role of Same-Sex Relations in Human Societies* (McFarland & Co., 2011)

Nishie, Hitonaru, and Michio Nakamura, "A newborn infant chimpanzee snatched and cannibalized immediately after birth: Implications for 'maternity leave' in wild chimpanzee," *American Journal of Physical Anthropology* 165: 194–99 (January 2018)

O'Connor, Sue, et al., "Fishing in life and death: Pleistocene fish-hooks from a burial context on Alor Island, Indonesia," *Antiquity* 91(360): 1451–68 (December 6, 2017)

Ólafsdóttir, H. Freyja, et al., "Hippocampal place cells construct reward related sequences through unexplored space," *Elife* 4: e06063 (June 26, 2015)

Olkowicz, Seweryn, et al., "Birds have primate-like numbers of neurons in the forebrain," *PNAS* 113(26): 7255–60 (June 28, 2016)

Organ, C., et al., "Phylogenetic rate shifts in feeding time during the evolution of *Homo*," PNAS 108(35): 14555–59 (August 30, 2011)

Powell, A., S. Shennan, and M. G. Thomas, "Late Pleistocene demography and the appearance of modern human behavior," *Science* 324(5932): 1298–1301 (June 5, 2009)

Prabhakar, Shyam, "Accelerated evolution of conserved noncoding sequences in humans," *Science* 314(5800): 786 (November 3, 2006)

Prabhakar, Shyam, et al., "Human-specific gain of function in a developmental enhancer," Science 321(5894): 1346–50 (September 5, 2008)

Pratt, D. M., and V. H. Anderson, "Population, distribution and behavior of giraffe in the Arusha National Park, Tanzania," *Journal of Natural History* 16(4): 481–89 (1982)

———, "Giraffe social behavior," *Journal of Natural History* 19(4): 771–81 (1985)

Prior, Helmut, et al., "Mirror-induced behavior in the magpie (*Pica pica*): Evidence of self-recognition," *PLoS Biology* 6(8): e202 (August 19, 2008)

Pruetz, Jill D., et al., "Savanna chimpanzees, *Pan troglodytes verus*, hunt with tools," *Current Biology* 17(5): 412–17 (March 6, 2007)

Pruetz, Jill D., and Nicole M. Herzog, "Savanna chimpanzees at Fongoli, Senegal, navigate a fire landscape," *Current Anthropology* 58(S16): S337–S350 (August 2017)

Pruetz, Jill D., and Thomas C. LaDuke, "Reaction to fire by savanna chimpanzees (*Pan troglodytes verus*) at Fongoli, Senegal: Conceptualization of 'fire behavior' and the case for a chimpanzee model," *American Journal of Physical Anthropology*141(14): 646–50 (April 2010)

Prüfer, Kay, et al., "The bonobo genome compared with the chimpanzee and human genomes," *Nature* 486: 527–31 (June 28, 2012)

Quiles, Anita, et al., "A high-precision chronological model for the decorated Upper Paleolithic cave of Chauvet-Pont d'Arc, Ardèche, France," *PNAS* 113(17): 4670–75 (April 26, 2016)

Rodríguez-Vidal, Joaquín, et al., "A rock engraving made by Neanderthals in Gibraltar," *PNAS* 111(37): 13301–06 (September 16, 2014)

Russell, Douglas G. D., et al., "Dr. George Murray Levick (1876–1956): Unpublished notes on the sexual habits of the Adélie penguin," *Polar Record* 48(4): 387–93 (January 2012)

Russon, Anne E., et al., "Orangutan fish eating, primate aquatic fauna eating, and their implications for the origins of ancestral hominin fish eating," *Journal of Human Evolution* 77: 50–63 (December 2014)

Ruxton, Graeme D., and Martin Stevens, "The evolutionary ecology of decorating ehavior," *Biology Letters* 11(6) (June 3, 2015)

Sahnouni, Mohamed, et al., "1.9-million- and 2.4-million-year-old artifacts and stone tool–cutmarked bones from Ain Boucherit, Algeria," *Science* 362(6420): 1297–1301 (December 14, 2018)

Saini, Angela, *Inferior: How Science Got Women Wrong* (Fourth Estate, 2017)

Sazima, Ivan, "Corpse bride irresistible: A dead female tegu lizard (*Salvator merianae*) courted by males for two days at an urban park in southeastern Brazil," *Herpetology Notes* 8: 15–18 (January 25, 2015)

Schnytzer, Y., et al., "Boxer crabs induce asexual reproduction of their associated sea anemones by splitting and intraspecific theft," *PeerJ* 5: e2954 (January 31, 2017)

Schmitz, Helmut, and Herbert Bousack, "Modelling a historic oil-tank fire allows an estimation of the sensitivity of the infrared receptors in pyrophilous *Melanophila* beetles," *PLoS ONE* 7(5): e37627 (May 21, 2012)

Scott, Erin M., et al., "Aggression in bottlenose dolphins: Evidence for sexual coercion, male–male competition, and female tolerance through analysis of tooth-rake marks and behaviour," *Behaviour* 142(1): 21–44 (January 2005)

Sergiel, Agnieszka, et al., "Fellatio in captive brown bears: Evidence of long-term effects of suckling deprivation?," *Zoo Biology* 9999: 1–4 (June 4, 2014)

Shakespeare, William, *The Tragedy of Hamlet, Prince of Denmark* (Folio 1, 1623)

Sporny, Michael, et al., "Structural history of human SRGAP2 proteins," *Molecular Biology and Evolution* 34(6): 1463–78 (June 1, 2017)

St Clair, James J. H., et al., "Hook innovation boosts foraging efficiency in tool-using crows," *Nature Ecology & Evolution* 2: 441–44 (January 22, 2018)

Steiner, A. P., and A. D. Redish, "Behavioral and neurophysiological correlates of regret in rat decision- making on a neuroeconomic task," *Nature Neuroscience* 17(7): 995–1002 (June 8, 2014)

Sudmant, Peter H., "Diversity of human copy number variation and multicopy genes," *Science* 330(6004): 641–46 (October 29, 2010)

Takemoto, Hiroyuki, et al., "How did bonobos come to range south of the Congo River? Reconsideration of the divergence of Pan paniscus from other Pan populations," *Evolutionary Anthropology* 24(5): 170–84 (September 2015)

Tan, Min, et al., "Fellatio by fruit bats prolongs copulation time," *PLoS ONE* 4(10): e7595 (October 28, 2009)

Taylor, Alex H., et al., "Spontaneous metatool use by New Caledonian crows," *Current Biology* 17(17): 1504–07 (September 4, 2007)

Thornhill, Randy, and Craig T. Palmer, *A Natural History of Rape: Biological Bases of Sexual Coercion*, (The MIT Press, 2000)

Trinajstic, K., et al., "Pelvic and reproductive structures in placoderms (stem gnathostomes)," *Biological Reviews* 90(2): 467–501 (May 2015)

Vargha-Khadem, Faraneh, et al., "Praxic and nonverbal cognitive deficits in a large family with a genetically transmitted speech and language disorder," *PNAS* 92(3): 930–33 (January 31, 1995)

Vernes, Sonja C., et al., "A functional genetic link between distinct developmental language disorders," *New England Journal of Medicine* 359: 2337–45 (November 27, 2008)

Visalberghi, Elisabetta, et al., "Selection of effective stone tools by wild bearded capuchin monkeys," *Current Biology* 19(3): 213–17 (February 10, 2009)

Waterman, Jane M., "The adaptive function of masturbation in a promiscuous African ground squirrel," *PLoS ONE* 5(9): e13060 (September 28, 2010)

White, Randall, "The women of Brassempouy: A century of research and interpretation," *Journal of Archaeological Method and Theory* 13(4): 250–303 (December 2006)

Wikelski, Martin, and Silke Bäurle, "Pre-copulatory ejaculation solves time constraints during copulations in marine iguanas," *Proceedings of the Royal Society B* 263: 1369 (April 2, 1996)

Wilson, Michael L., et al., "Lethal aggression in *Pan* is better explained by adaptive strategies than human impacts," *Nature* 513: 414–17 (September 18, 2014)

Zhu, Zhaoyu, et al., "Hominin occupation of the Chinese Loess Plateau since about 2.1 million years ago," *Nature* (July 11, 2018), doi.org/10.1038/s41586-018-0299-4

INDEX

ABOUT THE AUTHOR

ADAM RUTHERFORD is a geneticist, science writer, and broadcaster. He studied genetics at University College London, and during his PhD on the developing eye, he was part of a team that identified the first known genetic cause of a form of childhood blindness. As well as writing for the science pages of *The Guardian*, he has written and presented many award-winning series and programs for the BBC, including the flagship weekly Radio 4 program *Inside Science*, *The Cell* for BBC Four, and *Playing God* (on the rise of synthetic biology) for the leading science series *Horizon*. He is also the author of *How to Argue With a Racist*, an incisive guide to what modern genetics can and can't tell us about human difference; *A Brief History of Everyone Who Ever Lived*, finalist for the National Book Critics Circle Award in nonfiction; and *Creation*, on the origin of life and synthetic biology, which was short-listed for the Wellcome Book Prize.

@ADAMRUTHERFORD